AutoCAD® 2002:
Professional

AutoCAD® 2002:
Professional

BILL BURCHARD
DAVID PITZER

autodesk® Press

Australia • Canada • Mexico • Singapore • Spain • United Kingdom • United States

autodesk® Press

AutoCAD® 2002: Professional
Bill Burchard   David Pitzer

AutoDesk Press Staff

Business Unit Director:
Alar Elken

Executive Editor:
Sandy Clark

Acquisitions Editor:
James DeVoe

Development Editor:
John Fisher

Editorial Assistant:
Jasmine Hartman

Executive Marketing Manager:
Maura Theriault

Channel Manager:
Mary Johnson

Marketing Coordinator:
Karen Smith

Executive Production Manager:
Mary Ellen Black

Production Manager:
Larry Main

Production Editor:
Tom Stover

Art/Design Coordinator:
Mary Beth Vought

Production Services:
TIPS Technical Publishing

Library of Congress Cataloging-in-Publication Data
Pitzer, David.
   AutoCAD 2002: professional/
Dave Pitzer.
      p. cm.
   ISBN 0-7668-4369-6
   1. Computer graphics. 2. AutoCAD.
I. Title.

T385 .P527 2001
620'.0042'02855369--dc21
                              2001047330

CONTENTS

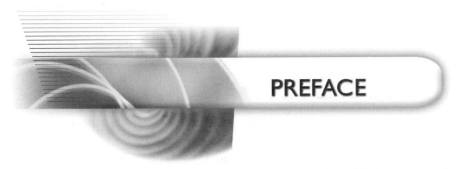

PREFACE

AutoCAD 2002 Professional is written for those who use AutoCAD on a regular basis, and who need to gain insight and knowledge on AutoCAD's advanced topics. Designed for intermediate AutoCAD users, this book gives you the power to master AutoCAD's more complex features and tools. Additionally, you learn about the challenges that office CAD managers face, and how experts resolve these CAD-related management issues.

WHY YOU NEED THIS BOOK

AutoCAD 2002 Professional provides an in-depth discussion of topics that you need to master to become an AutoCAD expert. This book covers topics such as customizing AutoCAD's interface and tools, creating powerful custom applications using Visual LISP and Visual Basic for Applications (VBA), and mastering AutoCAD's Internet-based tools and features. You also learn about AutoCAD management techniques that help you grow beyond simply being an AutoCAD user, to becoming an office or corporate-level CAD Manager.

AutoCAD 2002 Professional is written for experienced AutoCAD users who need to master the advanced tools and customization capabilities of the latest release of AutoCAD, and who want to gain insight into CAD management techniques from the experts.

WHAT YOU WILL FIND IN THIS BOOK

AutoCAD 2002 Professional provides an in-depth discussion of topics that you need to master to become an AutoCAD expert. This book provides concise, step-by-step explanations on how to use AutoCAD's advanced tools that let you work with OLE objects and external databases. You master the secrets of interconnecting files and documents from multiple applications to and from AutoCAD drawings. You learn about the power of connecting to external databases, and linking AutoCAD objects to records in tables, a capability that lets you automatically label objects in drawings using records from databases. You follow examples that show you how to use AutoCAD's database query tools to extract information from external databases, such as part numbers or lot sizes.

AutoCAD 2002 Professional helps you master AutoCAD's Internet-based tools and features. With this book, you learn how to publish your drawings to the Web, how to use the new i-drop feature, which lets you drag and drop content into your drawing from your Internet browser, and how to use the new Meet Now tool for collaborating on projects over the Internet.

AutoCAD 2002 Professional introduces you to the power of customizing AutoCAD. You learn about tools and features that let you customize AutoCAD's interface and commands without programming! You learn how to customize menus, and how to maximize the power of custom menus by using AutoCAD's DIESEL tool set. Through DIESEL, you add intelligence to AutoCAD's menu-driven interface. Additionally, you learn how the experts build powerful custom applications using AutoLISP, Visual LISP, and Visual Basic for Applications (VBA).

The Professional book gives you the insight you need to master AutoCAD management techniques that help you grow beyond simply being an AutoCAD user, to becoming an office or corporate-level CAD Manager. AutoCAD management issues are addressed, such as determining Return On Investment (ROI), and installing AutoCAD in the business environment. Through the ROI chapter, you learn techniques for calculating how long it takes your company to recoup their AutoCAD software and training investment, which is must-have knowledge for anticipating and overcoming your company's financial challenges. Once you purchase AutoCAD licenses, this book guides you through the process of installing AutoCAD in a networked environment. AutoCAD 2002 Professional also explains how to overcome other CAD management challenges, such as developing CAD Standards and managing CAD workflow.

HOW TO USE THIS BOOK

AutoCAD 2002 Professional is organized to help you get up-to-speed as quickly as possible by focusing on the information you need. The book is organized into five parts, with each part focusing on a specific AutoCAD theme. Part 1, Advanced Tools and Features, presents AutoCAD's advanced tools and helps you to achieve and maintain a high level of AutoCAD expertise. Part 2, CAD on the Internet, presents AutoCAD's Internet tools, and is where you master the power of AutoCAD's new Internet-based tools that let you publish drawings to the Web, and to collaborate on projects over the Internet. Part 3, Customizing AutoCAD, and Part 4, Developing Custom Applications, present customizing AutoCAD, where you learn to modify AutoCAD's interface and tools to maximize productivity, and to create your own custom applications. Part 5, Managing AutoCAD, presents solutions to the various challenges faced when managing AutoCAD, where you learn from experts who explain AutoCAD management techniques that cover topics such as developing CAD Standards, calculating return on investment (ROI), and managing CAD workflow.

CONVENTIONS USED IN THIS BOOK

Most topics covered in this book are detailed explanations of new features. Where needed for clarity, examples are presented that step you through the process of using a particular feature. Many of the examples use files located on the accompanying CD.

USING THE CD

All drawings, databases, and other files used in this book's examples are included on the accompanying CD. Typically, if an example uses a file from the CD, the example walks you through the process of copying the file to a new folder on you system, and then clearing the file's Read Only attribute.

ABOUT THE AUTHORS

Bill Burchard is Corporate CADD Manager for Psomas, a California-based land surveying and civil engineering firm. Mr. Burchard has served the AEC industry for 25 years. His range of experience includes surveying and engineering design, computer modeling and applications development, plan preparation and technical publications, as well as Web development, Geographic Information Systems (GIS), 3D modeling, 3D photo-realistic renderings, and 3D animations. Additionally, Mr. Burchard is a registered Autodesk author/publisher, and along with co-author David Pitzer, has written numerous books, including Inside AutoCAD 2000 Limited Edition, Inside AutoCAD 2000, Inside AutoCAD 14 Limited Edition, and Inside AutoCAD 14. Mr. Burchard also writes regularly featured articles for CADalyst magazine, published by Advanstar Communications, for Inside AutoCAD Journal, published by Element K Content, and for Autodesk's Toplines eNewsletter, which is accessed from Autodesk's Point A Web site (www.autodesk.com/pointa). He writes both feature articles and a bimonthly column for CADalyst and for Toplines, and writes monthly articles for Inside AutoCAD Journal. Additionally, Mr. Burchard sits on the Advisory Committee Board for Computer Sciences at Riverside Community College, and lectures on the subject of GIS for the University of California, Irvine.

David Pitzer has been using AutoCAD since 1987. He was formerly CADD Manager for an engineering firm in northern California where he lives. He has written numerous articles for CADENCE Magazine and CADalyst Magazine for whom he is a Contributing Editor. He is a member of the faculty at Santa Rosa Junior College where he teaches advanced AutoCAD courses. He has co-authored five AutoCAD books for New Riders Publishing. He has, in addition, written technical material for Autodesk.

ACKNOWLEDGMENTS

The authors wish to thank and acknowledge the many professionals who reviewed the manuscript, helping us publish this book. Thanks go to the Autodesk Press Team: Jim DeVoe, Acquisitions Editor, John Fisher, Developmental Editor, Mary

Beth Vought, Art & Design Coordinator, Stacy Masucci, Production Editor, Karen Smith, Marketing Coordinator, and Jasmine Hartman, Editorial Assistant. Thanks to Ed O'Halloran who performed the technical edit.

Special thanks is due to the following authors who contributed to this book:

Ralph Grabowski for authoring the Internet-related chapters in Part 2, CAD on the Internet.

John Gibb for authoring the chapter, "Introduction to VBA," in Part 4, Developing Custom Applications.

Mark Middlebrook for authoring the chapter, "Developing CAD Standards," in Part 5, Managing AutoCAD.

Bill Burchard wishes to thank his family and friends for their support during the last several months while writing this book. Additionally, Bill thanks his colleagues at Psomas for their support, suggestions, guidance, and help, with special thanks to Joanne Diaz, Sherri Bayer, and Amy McLaughlin, as well as his friends at Autodesk, including Glenn Cooper and Neill Vickers, with special thanks to Bob Ng, who contributed significantly to his chapter, "Determining ROI—Return on Investment." Finally, Bill thanks his fellow C3 Group members for tolerating his absence during the last several months while writing this book, with special thanks to Gene Redmon, President, and Devon Diaz, past President.

DEDICATIONS

Bill Burchard:

I dedicate this book to my nephews, Dominic and Tony Negron, Marines called to duty in our Country's time of need. They and their comrades are our guardians and our future heroes, and make me tremendously proud. I also dedicate this book to their father, Rudy, for inspiring two boys to become such fine young men.

Dave Pitzer:

I would like to dedicate this book to my students—past, present and future. Their excitement with AutoCAD is a constant source of enjoyment.

SECTION

I

AutoCAD's
Advanced
Features

Working with OLE Objects

When you work on a set of drawings, you are typically working on one element of a project. Other elements may include text documents, spreadsheet data, and graphics created in programs other than AutoCAD. All of these elements combined are required to complete the project's deliverables and to meet your clients' needs.

Quite often, data created in other applications must be duplicated in your AutoCAD drawing. Elements such as general notes created in a word processing program, or a Bill of Materials created in a spreadsheet application, must be duplicated in your drawing to satisfy the project's final delivery requirements. By adding this data to your drawing, you make the drawing a complete project deliverable.

Developing compound documents using Object Linking and Embedding (OLE) is a powerful, simple way to create the final documents required to satisfy your clients' needs. By inserting documents created using other applications into your AutoCAD drawing, you create a compound document. By simply dragging existing files into your drawing, you can insert data created using word processing applications such as Word or WordPerfect, as well as tabular data from spreadsheet programs such as Excel or Lotus, directly into your AutoCAD drawing. By using OLE, you make the process of completing a set of drawings easier because you use existing data in its native format.

This chapter reviews AutoCAD OLE capabilities and covers the following subjects:

- Understanding OLE
- Importing objects into AutoCAD using OLE
- Exporting AutoCAD Objects using OLE

UNDERSTANDING OBJECT LINKING AND EMBEDDING

Object Linking and Embedding (OLE) is a feature provided by the Windows operating system. Whether or not an application takes advantage of OLE is up to its program developers. In the case of AutoCAD, the application is designed to take advantage of OLE technology, allowing you to interact with other OLE-compliant applications. When you use OLE, you can insert files from other applications directly into AutoCAD drawings, and you can insert AutoCAD views and AutoCAD objects into other OLE-compliant applications.

Object linking and embedding refers to the two different ways you can insert a file from another application into your drawing. You can insert an OLE object as a linked object or as an embedded object. A linked object inserts a copy of a file, and the copy references the original source file. A linked OLE object behaves similarly to xrefs in that any modifications made to the source file are reflected in the linked OLE object when the link is updated in your drawing.

In contrast, while an embedded object also inserts a copy of a file into your drawing, it does not maintain a link to the source file. An embedded OLE object behaves similarly to a block inserted from another drawing in that the inserted file exists independently of the source from which it was copied, and it may be edited independently without affecting the source file. More importantly, edits made to the source file are never reflected in the embedded OLE object. Use linked objects when you want modifications to the source file to be displayed in your drawing, and use embedded objects when you want to insert a copy of a file and do not want edits to the source file to be displayed in your drawing.

OLE objects inserted into AutoCAD drawings have certain limitations. For example, you can only insert one page of a word processor's document at a time. Also you can only insert a portion of a spreadsheet file: a limited number of rows and columns amounting to an area approximately 10 inches wide and 13 inches long. Another limitation is that OLE objects cannot be resized if they are rotated in your drawing. Even with these limitations, you will find object linking and embedding a very useful feature.

Note: Windows, not AutoCAD, imposes the limitations of OLE.

Tip: You can use the OLESTARTUP system variable to optimize the quality of plotted OLE objects. The variable controls whether the source application of an inserted OLE object loads when AutoCAD plots. Setting the value to one instructs AutoCAD to load the OLE source application when plotting. Setting the value to zero instructs AutoCAD not to load the OLE source application when plotting.

IMPORTING OBJECTS INTO AUTOCAD USING OLE

You can create compound documents in AutoCAD by linking or embedding objects from other applications. For example, you can insert a table from a spreadsheet application, a set of notes from a word processing application, and a graphic image from a paint program. By inserting the desired objects into your AutoCAD drawing, you create a compound document.

AutoCAD provides several options for linking and embedding objects in drawings, as described in the following sections.

INSERTING OLE OBJECTS FROM WITHIN AUTOCAD

You can insert OLE objects into AutoCAD using the Insert Object dialog box. This procedure allows you to insert linked or embedded objects from within AutoCAD by executing an AutoCAD command. From the Insert Object dialog box you can insert an object from an existing file or create a new OLE object that exists only in the current drawing.

The Insert Object dialog box is opened from the Insert menu by choosing OLE Object. Once opened, the Insert Object dialog box presents a list of object types that it can link or embed, as shown in Figure 1–1.

From the Insert Object dialog box you select whether you want to create a new OLE object or insert an OLE object from an existing file. The Create New option opens the selected application so you can create the object. Then, when the object is saved, the selected application closes, and AutoCAD embeds the object in the current drawing.

In contrast, when you choose the Create from File option, the Insert Object dialog box changes its display, allowing you to browse for an object to link or embed, as shown in Figure 1–2. When you select the Link check box, the selected object is inserted into AutoCAD and linked back to the original file.

The Insert Object dialog box provides a straightforward method for inserting OLE objects. Because it gives you the option of either creating new OLE objects or browsing for existing object files, you can easily insert the needed OLE object into your AutoCAD drawing.

PASTING OLE OBJECTS INTO AUTOCAD

You can insert OLE objects into AutoCAD by pasting them from the Windows Clipboard. This procedure is a very common way to insert OLE objects from one

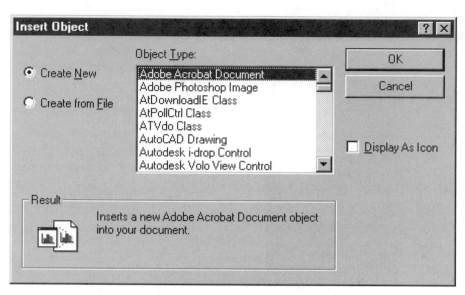

Figure 1–1 *The Insert Object dialog box allows you insert OLE objects from within AutoCAD.*

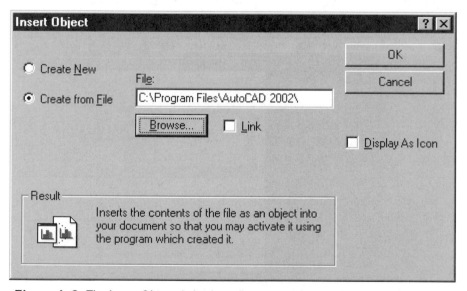

Figure 1–2 *The Insert Object dialog box allows you to browse for existing OLE object files to insert into AutoCAD.*

application to another. Using this feature you can copy an object directly from its application to the Clipboard and then paste the Clipboard's contents into AutoCAD.

You paste objects from the Clipboard using either the Paste command or the Paste Special command. These commands are accessed from AutoCAD's Edit menu. The Paste command is also accessed from the shortcut menu, which is displayed by right-clicking in the drawing area. These commands are only available when the Clipboard contains objects.

 Note: You can view the contents of the Clipboard using the Clipboard Viewer, and you can also delete the Clipboard's contents. You access the Clipboard Viewer from the Windows Taskbar by choosing Start>Programs>Accessories>Clipboard Viewer. You view the Clipboard's contents by opening the Clipboard window, which appears as an icon in the Clipboard Viewer. To delete the Clipboard's contents, from the Clipboard Viewer's Edit menu, choose Delete.

While both commands paste objects into the current drawing from the Clipboard, they differ in one important way. The Paste command only embeds objects. The Paste Special command allows you either to embed objects or insert them as linked objects.

When you choose the Paste command, the object is immediately embedded into AutoCAD. Additionally, the OLE Properties dialog box is displayed if the Display Dialog Box check box is selected, when Pasting New OLE Objects . The OLE Properties dialog box, which is discussed in detail later in this chapter, allows you to control the size of the OLE object.

When you choose the Paste Special command, AutoCAD displays the Paste Special dialog box. From this dialog box you can choose either the Paste option or the Paste Link option.

When you use the Paste option, the OLE object is embedded into the drawing. The difference between pasting an object from the Paste Special dialog box versus pasting it directly from the Edit or shortcut menus is that when you use the Paste Special dialog box, you have more control over the OLE object type you are embedding.

When you choose the Paste option in the Paste Special dialog box, the available object types are displayed in the As list. The object types listed depend on the OLE object you are pasting from the Clipboard. For example, if the Clipboard contains a Microsoft Word document, you can embed the Clipboard's contents as one of several object types shown in Figure 1–3. The list only displays acceptable types for the particular object. Several object types are listed and described as follows:

- **Picture (Metafile)**—Inserts the contents of the Clipboard into your drawing as a vector-based picture.

- **AutoCAD Entities**—Inserts the contents of the Clipboard into your drawing as circles, arcs, lines, and polylines. Text is inserted as text objects, and each line of text located in a paragraph in the source file is converted to an individual AutoCAD text object.

- **Image Entity**—Inserts the contents of the Clipboard into your drawing as an AutoCAD raster image object.

- **Text**—Inserts the contents of the Clipboard into your drawing as an AutoCAD MTEXT object. Any line objects are ignored.

- **Package**—Inserts the contents of the Clipboard into your drawing as a Windows Package object. A package is an icon that represents embedded or linked information. The information may consist of a complete document, such as a Paint bitmap, or part of a document, such as a spreadsheet cell. You create packages using the Windows Object Packager, which is accessed from the Taskbar by choosing Start>Programs>Accessories>Object Packager.

- **Bitmap Image**—Inserts the contents of the Clipboard into your drawing as a bitmap image object.

When you choose the Paste Link option, you can only insert the OLE object as its original object type. For example, if you choose the Paste Link option to insert a Microsoft Word document, you can only insert it as a Microsoft Word document. This ensures that you can open and edit the source Word document to which the OLE object is linked.

Tip: If you work in a black background in AutoCAD and paste an image from a word processing or spreadsheet application, the pasted image will appear with a white background in the drawing. If this is undesirable, change your background in either program to match the other.

By using the Paste and Paste Special commands, you can embed or link OLE objects in your AutoCAD drawing from the Clipboard. Next you will learn about another method for inserting OLE objects.

USING DRAG & DROP TO INSERT OLE OBJECTS

The Windows operating system provides the ability to drag and drop selected objects from another application into an AutoCAD drawing. By selecting objects in an open application, then dragging the selected objects into AutoCAD, you, in effect, cut the objects from the application and embed them into the AutoCAD drawing. To copy the objects from the open application, instead of cutting them, hold down the Ctrl key while dragging the objects into the drawing.

Additionally, you can drag and drop objects from Windows Explorer. If AutoCAD recognizes the object type, it embeds the object into the drawing. If AutoCAD does not recognize the file type, or if the object cannot be inserted as an OLE object, AutoCAD issues an error and cancels the function.

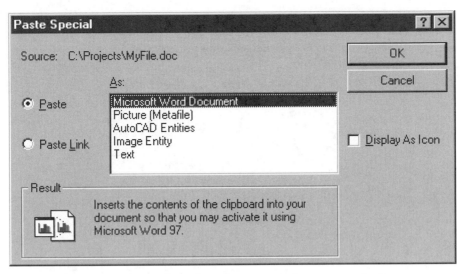

Figure 1–3 *The Paste Special command allows you to select an object's type when it is embedded.*

 Tip: After you select objects in an application, drag them into AutoCAD by right-clicking with your pointing device. When you release, AutoCAD displays the shortcut menu and allows you either to move the objects (Cut & Paste), copy the objects (Copy & Paste), or link the objects (Paste Special) into the AutoCAD drawing. You can also create a hyperlink to the file or cancel the operation from the shortcut menu.

CONTROLLING OLE OBJECT PROPERTIES

AutoCAD provides specific tools for manipulating an OLE Object, because common AutoCAD commands typically do not affect OLE objects. For example, you cannot select an OLE object and erase it with the ERASE command, nor can you resize it using the SCALE command. However, by using specific tools designed for manipulating OLE objects, you can control an OLE object's appearance in AutoCAD.

 Tip: Once you select an OLE object in AutoCAD to display its grips, you can press the Delete key to delete the object.

CONTROLLING OLE OBJECT SIZE

AutoCAD lets you control an OLE object's size in a drawing through the OLE Properties dialog box. The dialog box allows you to control an object's size in one of three ways, as shown in Figure 1–4. You access the dialog box by right-clicking over an OLE object and selecting Properties from the shortcut menu.

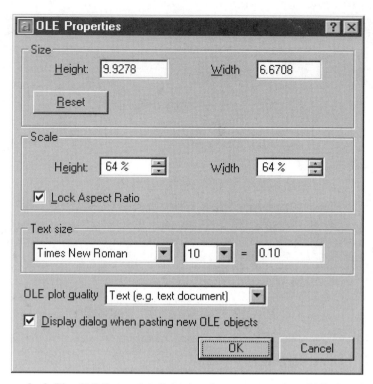

Figure 1–4 *The OLE Properties dialog box lets you control an OLE object's size.*

In the Size area, you control an OLE object's size by entering values in the Height and Width fields. If the Lock Aspect Ratio check box is selected, when one field value is changed, the other gets updated automatically, proportionally maintaining the OLE object's aspect-ratio size. The units entered in the field are based on the drawing's current units setting. You can also set the OLE object back to its original size by choosing the Reset button.

In the Scale area, you control the OLE object's size by entering a percentage of the object's size. As with the values in the Size area, if the Lock Aspect Ratio check box is selected, when one field value is changed in the Height or Width fields, the other value gets updated automatically.

A third method for controlling an OLE object's size is available in the Text Size area. If the OLE object contains text, you can enter a new text size value to adjust the object's size. The first field displays the font styles in the OLE object, and the second field contains a list of the selected font's sizes. Once you choose the desired font style and size in the first two fields, you can control the object's overall size by entering the desired height of the text in the third field. For example, in Figure 1–4, the OLE object will be resized based on the 10 point Times New Roman font being set to 0.10 drawing units.

 Note: It is important to understand that the three areas provided for controlling an OLE object's size work in unison. When one set of values is changed in one area, the values are automatically changed in the other two areas. The values in the three areas cannot be set independently of each other.

In addition to controlling an OLE object's size, the OLE Properties dialog box provides the ability to control the plot quality of an OLE object. From the OLE plot quality list you can choose one of five plot quality options:

- Line Art—Intended for plotting objects such as a spreadsheet
- Text—Intended for plotting objects such as a Word document
- Graphics—Intended for plotting objects such as a pie chart
- Photograph—Intended for plotting objects that are color images
- High Quality Photograph—Intended for plotting objects that are true-color images

The plot quality options are applied specifically to the selected OLE object. Therefore, you can insert a Word document that contains only text and then set the plot quality to Text. You can then insert a true-color image and set its plot quality to High Quality Photograph. By applying the desired plot quality to each OLE object, you can control an object's appearance when it is plotted.

CONTROLLING OLE OBJECTS USING THE SHORTCUT MENU

Once an OLE object is inserted into a drawing, you can control several object properties and perform edits through commands accessed from the shortcut menu. By using these commands, you can delete the OLE object or copy it to the Clipboard. You can determine if the object appears on top or below other objects in the drawing, and you can control whether or not it may be selected for editing. The shortcut menu offers these commands and more, providing useful control over OLE objects.

Cutting, Copying, and Clearing OLE Objects

When you right-click over an OLE object, AutoCAD displays the available OLE shortcut commands, as shown in Figure 1–5. The first three commands do the following:

- **Cut**—Erases the selected object from the drawing and places a copy in the Clipboard. You can also execute the Cut command by pressing Ctrl+X.
- **Copy**—Leaves the selected object in the drawing and places a copy in the Clipboard. You can also execute the Copy command by pressing Ctrl+C.

- **Clear**—Erases the selected object from the drawing without placing a copy in the Clipboard. You can also execute the Clear command by entering E and then pressing Enter.

When you use the Cut or Copy commands on an OLE object, the object is placed on the Clipboard in its original object format, not as an AutoCAD object. For example, suppose a Word document resides in your drawing as an OLE object. If you use the Copy command to copy the Word document object from your drawing to the Clipboard, when you paste the object into another application, it is pasted as a Word document object.

Undoing OLE Object Edits

The next command on the shortcut menu is Undo, which undoes edits made to the OLE object while it is in your drawing. For example, if you move or resize the object in your drawing, you can undo the edit by selecting Undo from the shortcut menu. By selecting the Undo command repeatedly, you can undo a series of edits made to the object.

The Undo command has one important limitation: It does not undo edits made to the OLE object in the object's source application. For example, if you paste a Word document object into your drawing and then edit the document object in Word by adding additional text, when you save your edits and return to your drawing, the edits made in Word cannot be undone in AutoCAD. In other words, the text added to the document from the Word application cannot be undone using the Undo command in AutoCAD.

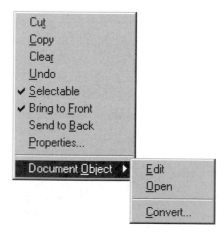

Figure 1–5 *Right-click over an OLE object to display the shortcut menu commands.*

Note: Do not confuse the Undo command found on the OLE object shortcut menu with AutoCAD's UNDO command. The UNDO command (Ctrl+Z), which is accessed from AutoCAD's Edit menu, does not undo edits, such as moving or resizing the object, made to an OLE object while it is in your drawing. However, AutoCAD's UNDO command will undo the command used to paste the object into your drawing, thereby removing the entire object from your drawing.

Controlling OLE Object Selectability

The next item on the shortcut menu is the Selectable property, which toggles the selectability of the OLE object. When this property is toggled on, a check appears next to the property, indicating that the OLE object may be selected and then moved or resized in your drawing. When toggled off, the check is cleared, indicating that the object cannot be selected.

When you select an OLE object whose Selectable property is toggled on, an object frame and its sizing handles appear around the object. When you place your cursor inside the object frame, the cursor changes to the Move cursor, which is an icon comprised of a cross with four arrows, as shown in Figure 1–6. The Move cursor allows you to drag the object to a new position in your drawing.

The sizing handles are the small, solid squares that appear at the corners and midpoints of the object frame. When you place your cursor over a sizing handle, the cursor changes to a double-headed arrow. You can then resize the object by dragging the sizing handle. The sizing handles at the midpoints of the object frame stretch the object, distorting its appearance. The sizing handles at the corners of the object frame scale the object proportionally, maintaining the object's aspect ratio.

Tip: The sizing square color is controlled by the unselected grip color. This can be changed from the Options command dialog, Selection tab in the Grips area.

The three objects in Figure 1–7 are OLE objects inserted into a drawing from a Word document, and they provide an example of how stretching and resizing affects an object. The top object shows how all three objects appeared in their original size. The middle object is a copy of the top object that was stretched by dragging its sizing handles at the midpoints on each side of the object frame. Notice that the height of the text is the same as in the original. Only the width of the text is changed. The bottom object is also a copy of the top object, and it was resized by dragging its sizing handles in the corners of the object frame. Notice that while the text's height and width are larger than the original, their aspect ratio is maintained, and their overall size is correctly proportioned.

When the Selectable property is toggled off, you cannot select the object to display its object frame and sizing handles. This feature is useful for maintaining an object's

This is a Word document inserted as an OLE object.

Figure 1–6 *AutoCAD lets you drag an OLE object to a new position.*

Object's original size

O b j e c t   i s   s t r e t c h e d

Object is resized

Figure 1–7 *The original object is at the top, the stretched object is in the middle, and the resized object is at the bottom.*

 Tip: To quickly duplicate an OLE object inserted into AutoCAD, select the object and then drag it while pressing the Ctrl key.

size and position in your drawing once it has been set as desired. By clearing an OLE object's Selectable property, you ensure that the object will not be moved or resized in your drawing.

It is important to note that the Selectable property does not affect the other commands and properties on the shortcut menu. For example, while you cannot drag an OLE object's sizing handles when its Selectable property is cleared, you can resize the object through the OLE Properties dialog box. By right-clicking over the OLE object and choosing the Properties command from the shortcut menu, you display the OLE Properties dialog box. Any changes made to the object's size in the dialog box modify the object, even though its Selectable property is cleared. This is true for the shortcut menu's other commands, including Cut, Copy, Clear, and Undo.

Controlling OLE Object Display Order

The next two commands on the shortcut menu are Bring to Front, and Send to Back. These two properties control an object's position in the drawing relative to other objects, and they perform the same function as AutoCAD's Display Order tools. By right-clicking your cursor over an OLE object, then selecting Bring to Front or Send to Back from the shortcut menu, you place the object above or below other objects, as shown in Figure 1–8. These two properties are actually a toggle, and AutoCAD allows you to set only one for each OLE object.

 Note: Clicking inside an OLE object always selects it, even though it may be behind other AutoCAD objects.

 Tip: To select AutoCAD objects that lie within an OLE object's frame, clear the object's Selectable property.

Editing and Converting OLE Objects

At the bottom of the shortcut menu is the Object menu item. When selected, the Object item menu displays a cascading menu that provides access to the OLE object's source application through the Edit and Open commands, allowing you to modify the object. Additionally, you can access the Convert dialog box, which specifies a different source application for the OLE object. From the Object item menu, you can modify an OLE object or convert it to a different object type.

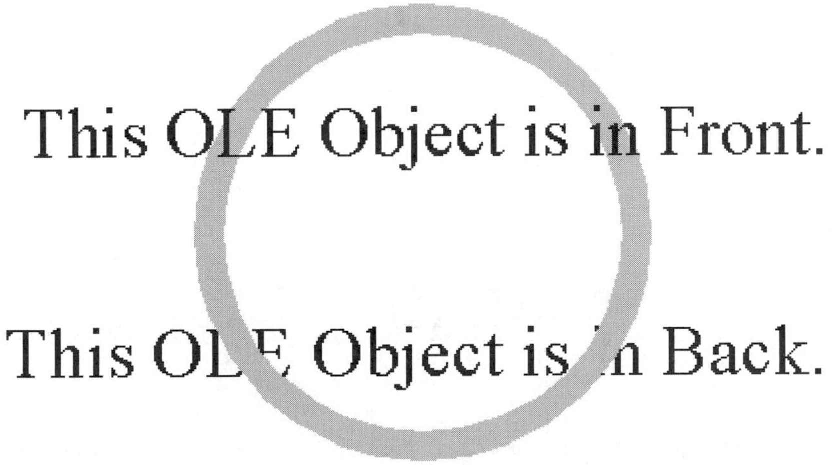

This OLE Object is in Front.

This OLE Object is in Back.

Figure 1–8 *The OLE object at the top has its Bring to Front property set, while the object at the bottom has its Send to Back property set.*

The name displayed for the Object item menu in the shortcut menu changes depending on the type of OLE object selected when the menu is accessed. For example, the shortcut menu in Figure 1–5 contains the menu item Document Object, which indicates that the object is a document object, specifically a Microsoft Word document object. If a Microsoft Excel worksheet is inserted as an OLE object and the shortcut menu is accessed, the menu item appears as Worksheet Object, as shown in Figure 1–9.

The Edit and Open commands launch the object's source application, displaying the object's source file and allowing you to make modifications. In AutoCAD both commands perform the same function. The reason the two commands perform the same function has to do with how AutoCAD interfaces with the Windows operating system when dealing with OLE objects.

In some applications other than AutoCAD, the Edit command only opens a source application file window inside the current application, while the Open command launches the entire source application. For example, Figure 1–10 shows what happens when you modify a Word document object inserted in an Excel spreadsheet by accessing the Edit command. Notice that the Word document window is inside the Excel spreadsheet. This allows you to make edits to the Word document from inside Excel without actually launching the Word application. In AutoCAD the Edit command launches the entire source application. Therefore, choosing either the Edit command or the Open command launches the source application, allowing you to make modifications.

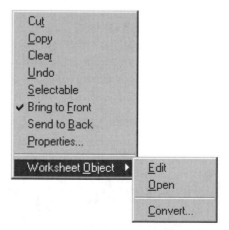

Figure 1–9 *The Object menu item description at the bottom of the shortcut menu changes to indicate that a Microsoft Excel worksheet is selected.*

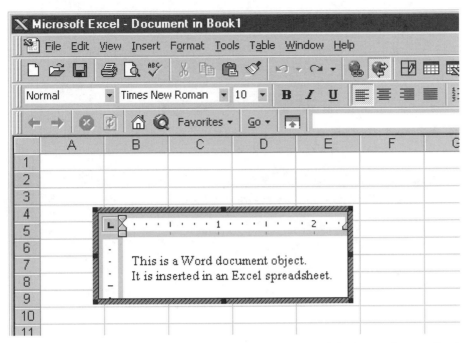

Figure 1–10 *The Edit command allows you to modify a Word document object in Excel.*

 Tip: You can execute the Edit command by double-clicking inside an object's frame.

The Convert command opens the Convert dialog box, which allows you to specify a different source application for an embedded object. When you select the desired source application and then choose OK, the object's source application type is changed to the new application type.

The different object types to which you can convert an object depend on the object selected. For example, when you select an embedded Word document object and then choose the Convert command, you are allowed to convert the document object to a Word picture object, as shown in Figure 1–11. The object types listed are those supported by the source application.

When you convert an object, you can choose from one of two options: Convert To and Activate As. The Convert To option converts an embedded object to the type specified under Object Type. This means that the object is actually converted to the new selected object type. For example, if you converted a Word document object into a Word picture object and then right-clicked over the object, the shortcut menu

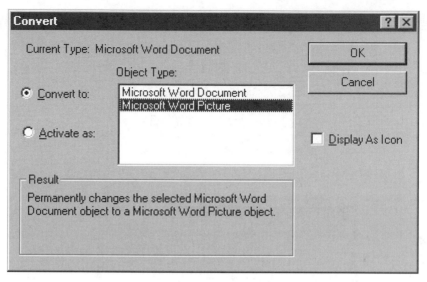

Figure 1–11 *The Convert dialog box allows you to convert the selected object to a different object type.*

would list the object as a Picture object. This means that when you edited the object, you would edit it as a Word picture not as a Word document.

The Activate As option acts similarly to the Convert To option, except it only temporarily converts the object to the selected object type during the editing process. Once the editing is complete, the object returns to its original type. For example, you can edit a Word document object, temporarily activating it as a Word picture object. This means the document object opens as a picture object in the Word application, which allows you to modify the object using Word's picture editing tools. Once you finish modifying the picture and close and save the file, the modified object is displayed in AutoCAD in its original format as a Word document object.

If you look at Figure 1–11 you will notice that you can also convert a selected Word document object to a Word document object. This appears to be a needless option and is a little confusing. This option is intended to maintain an object in its current type in conjunction with the Display As Icon check box. For example, when you choose to convert the Word document object into a Word document object, and then choose the Display As Icon option, the document object maintains its current type and changes its appearance so it is displayed as a Word document icon, as shown in Figure 1–12. To return the document object back to its original display, convert the Word document object into a Word document object, and then clear the check box.

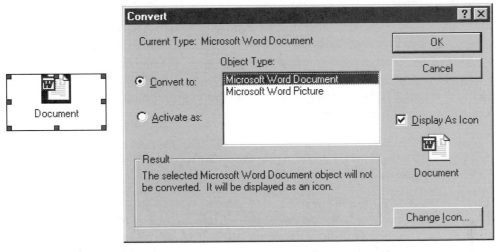

Figure 1–12 *The Convert dialog box allows you to display an OLE object as an icon.*

 Tip: Double-clicking on an object's icon launches the object's source application and allows you to modify the file.

 Note: When you select the Display As Icon option, the Change Icon button is activated. When selected, this button displays the Change Icon dialog box, which allows you to select a new icon to display as the object icon.

The OLE object shortcut menu provides several useful commands and options that allow you to control an OLE object's appearance and its behavior. Next you will learn about two features that allow you to control an OLE object's visibility.

CONTROLLING OLE OBJECT VISIBILITY

When you insert OLE objects into your drawing, you may want to control their visibility. Whether objects are inserted only temporarily or used for drawing-construction purposes, the way you might use construction lines or rays, you may want the visibility of OLE objects turned off in your drawing or in plotted sheets. By controlling OLE object visibility, you can use the objects to assist you in your work (by displaying reference information), and you can control when the objects are visible in your drawings and whether they appear in plots.

AutoCAD allows you to control the visibility of OLE objects through two methods. The first is simply to insert the OLE object on a layer that you turn off or on or that you freeze or thaw. The second is to use a special command that allows you to control globally the visibility of all OLE objects. By using these two features, you can easily control the visibility of OLE objects inserted in your drawing.

Controlling OLE Object Layer Properties

Controlling object visibility from the layer on which an object is inserted is a very common method for controlling whether or not an object is displayed in your drawing. One of the chief reasons to use layers to organize the objects in your drawing is to control the visibility of groups of objects that reside on a common layer. By inserting OLE objects on their own layers, you can easily control their visibility from the Layer Properties Manager.

 Tip: To move or copy an OLE object to a new layer, you must cut or copy the object to the Clipboard, make current the layer to which you want to move or copy the object, and then paste the object.

 Note: When you cut or copy an OLE object from an AutoCAD drawing and then paste it back into the drawing, the object's size will revert to its original size, and any modifications to the object will be lost. Therefore, the OLE object's size must be reset to the desired value after it is pasted back into the drawing. You can resize the OLE object from the Object Properties dialog box.

The Layer Properties Manager allows you to control more than just an OLE object's visibility. Specifically, OLE objects react to the following layer properties:

- **On/Off and Freeze/Thaw**—The On/Off and Freeze/Thaw layer properties control an OLE object's visibility, both on-screen and when plotted. Turning off or freezing the layer on which an OLE object resides no longer displays the object in your drawing, either on-screen or when plotted. To restore the object's visibility, turn on or thaw the layer.

- **Lock**—The Lock property prevents the OLE object from being selected. It functions similarly to the OLE object's Selectable property, except it does not allow any type of edits from the OLE object shortcut menu. For example, if an OLE object's Selectable option is turned off, you can still edit the object using commands from the shortcut menu, such as Properties, Edit, and Open. In contrast, when the layer on which an OLE object is inserted is locked, the OLE object shortcut menu cannot be invoked. The Lock property absolutely prevents the OLE object from being edited.

- **Plot**—The Plot property allows an object to remain visible on-screen but prevents it from plotting. This feature is useful if you need to display an OLE object during an editing session for reference information only but you do not want the object to appear when a drawing is plotted. By turning off the Plot property, you ensure that the OLE object will not appear on plotted drawings.

By using the layer properties discussed in this section, you can control both the appearance and the behavior of an OLE object through the layer on which it resides. In the next section you will learn how to control OLE object visibility globally.

Globally Controlling OLE Object Visibility

AutoCAD provides a method to control OLE object visibility globally. The OLE-HIDE command allows you to determine if OLE objects are visible in a drawing. When you use the OLEHIDE command, you control the visibility of all OLE objects and control whether they display in paper space or in model space.

The OLEHIDE command is actually a system variable whose current setting is stored in your computer's system registry. This means that when you set a value for OLEHIDE, the setting affects all drawings in the current editing session as well as in future sessions. To control the display of all OLE objects in all drawings, set the desired display value for the OLEHIDE system variable.

Typing OLEHIDE at the command prompt allows you to set the current OLE-HIDE system variable value. There are four possible integer values you can set:

0—Makes all OLE objects visible, both in paper space and in model space

1—Makes OLE objects visible only in paper space

2—Makes OLE objects visible only in model space

3—Makes all OLE objects invisible, both in paper space and in model space

When you set these values you control the appearance of all OLE objects, both in paper space and in model space.

Note: If the OLEHIDE system variable is set to 1 or 2 when you insert a new OLE object, AutoCAD automatically changes the OLEHIDE system variable value to allow the new OLE object to be displayed in the current space. This will also cause all OLE objects in the current space to appear.

In the next section you will learn how to work with OLE objects that are linked to their original file.

WORKING WITH LINKED OLE OBJECTS

When you insert an OLE object and link it to its original file, editing the file in its source application will automatically update the linked object. This means that when a linked object is inserted in a drawing and the object's original file is modified, the linked OLE object in AutoCAD is automatically updated to reflect the modifications. This feature is very useful for ensuring that the latest version of an inserted OLE object is displayed in an AutoCAD drawing.

While the ability to update and display the latest version of a linked file automatically is very useful, there will probably be occasions when you do not want the linked object to be automatically updated. For example, if you want to save permanently a set of drawings that represent a 50%-completed set, then you do not want OLE objects to be automatically updated when the 50%-completed set of drawings is re-opened for reference in the future. You need the ability to control whether or not linked OLE objects are automatically updated.

AutoCAD provides a tool that allows you to control whether or not linked OLE objects are automatically updated. When you choose the OLE Links command from the Edit menu, the Links dialog box is displayed, as shown in Figure 1–13. From the Links dialog box you can choose either the Automatic or Manual Update options, which control whether linked objects are automatically or manually updated. Additionally, from the Links dialog box you can restore links lost because the original file cannot be found, and you can associate the link to a different file. You can also break the link connection between the OLE object and the original file, which converts the linked object to an embedded object. From the Links dialog box you control the link between an OLE object and the file to which it is linked.

In the Links dialog box, if you choose the Manual Update option, you may update the link by choosing the Update Now button, also on the Links dialog box. The Open Source button opens the linked file's source application, allowing you to edit the source file. The Change Source button allows you either to locate a missing source file or select a new file. The Break Link button terminates the link between the OLE object and the source file. The link cannot be reestablished; once the link is broken, the OLE object is permanently detached from its original source file.

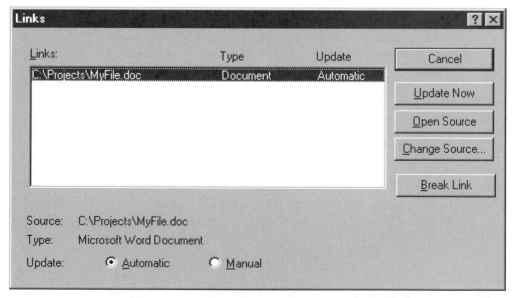

Figure 1–13 *The Links dialog box lets you control the link between an OLE object and its source file.*

EXPORTING AUTOCAD OBJECTS USING OLE

Just as you can insert files from other applications into AutoCAD, you can also insert AutoCAD drawings into other application files. By using certain commands created specifically for AutoCAD drawings and AutoCAD objects, you can insert either linked or embedded AutoCAD files into other application files. Therefore, AutoCAD's OLE features are designed for inserting files into AutoCAD as well as for inserting AutoCAD drawings into other applications.

Exporting AutoCAD objects into other application files as OLE objects involves determining if the AutoCAD objects will be linked or embedded. Linked objects are based on a named view in AutoCAD. When the view is updated in the AutoCAD drawing, the link is updated and the modified view appears in the application's file. In contrast, embedded objects are AutoCAD objects selected in the drawing and then copied to the Clipboard. Once pasted from the Clipboard, the objects are inserted as independent objects with no association to the original AutoCAD objects. Therefore, if the original AutoCAD objects are edited in the drawing from which they were copied, the objects embedded in the application will not be updated.

AutoCAD provides three commands for exporting AutoCAD information into other applications for linking and embedding:

- **Cut**—Executes the CUTCLIP command, which copies AutoCAD objects to the Clipboard, erasing the selected objects from the drawing

- **Copy**—Executes the COPYCLIP command, which copies AutoCAD objects to the Clipboard

- **Copy Link**—Executes the COPYLINK command, which copies the current AutoCAD view to the Clipboard

All three commands are located on AutoCAD's Edit menu.

When you use the Cut or Copy commands, AutoCAD prompts you to select objects if no objects are currently selected. Once the objects are selected, AutoCAD copies the selected objects to the Clipboard. If objects are selected prior to executing the commands, the selected objects are immediately copied to the Clipboard, and the command ends. If the Copy Link command is selected, AutoCAD copies all objects in the current view to the Clipboard, without prompting you for object selection. Therefore, the main difference between the Cut and Copy commands and the Copy Link command is that the Cut and Copy commands prompt you to select objects, while the Copy Link command does not.

When the AutoCAD objects are pasted into the target application, an object frame surrounds the objects and represents the drawing's viewport display at the time the objects were copied. This is true for all three commands. Therefore, whether you use the Cut or Copy commands to select objects or use the Copy Link command to automatically select objects, the AutoCAD OLE object pasted into the target application includes the visible area displayed in the current viewport.

When you paste an AutoCAD OLE object that was copied using the Copy Link command, AutoCAD creates a named view representing the current viewport display. This is necessary to maintain the link and accurately update the OLE object when the drawing file is modified. Associating the OLE object with a named view causes modifications to AutoCAD objects in the area of the drawing defined by the named view to automatically update the display of these objects in the target application.

 Tip: You can use the Copy Link command to paste an existing named view to the Clipboard by setting current the named view immediately before executing the Copy Link command.

 Tip: Often the main complaint with importing drawings into other applications is the Lineweight control. In the past it was difficult to have weights that accurately reflected a plotted drawing. However, starting with AutoCAD 2000, the use of lineweights lets you create drawings that display line widths when they are inserted as OLE objects in other applications.

SUMMARY

In this chapter you reviewed how to insert OLE objects from within AutoCAD, how to paste OLE objects into AutoCAD from the Clipboard, and how to use Drag & Drop to insert OLE objects into AutoCAD drawings. You learned how to control various OLE object properties, including how to resize an OLE object and how to control its visibility. You also reviewed how to edit OLE objects and how the Layer Properties Manager can affect certain OLE object properties. Finally, you learned how to export objects from AutoCAD into other applications, creating AutoCAD OLE objects.

CHAPTER 2

Working with External Databases

External databases are files that store information. The files are typically composed of tables that look similar to a spreadsheet, with data organized in columns called *fields*, and each unique set of data stored in rows called *records*. By using external databases to organize and store data, you can create, edit, and retrieve tremendous volumes of data.

AutoCAD 2002 provides tools that let you work with external database files. You can open database tables and view or edit their data. You can link database records to AutoCAD objects such as lines, circles, and polylines. You can insert data as labels (text objects) in your drawing that are automatically updated as data records change. You can run *queries*, which retrieve a subset of records based on certain criteria. By using AutoCAD's tools, you can access and use external database files from entirely within AutoCAD.

This chapter reviews AutoCAD's database tools and explains the steps necessary to work with external database files from directly within AutoCAD. This chapter covers the following subjects:

- Setting up AutoCAD to work with external databases
- Working with database tables
- Working with data and objects
- Using queries

SETTING UP AUTOCAD TO WORK WITH EXTERNAL DATABASES

The first step in using external databases with AutoCAD 2002 drawings requires defining information about the database files that you are using. Before AutoCAD (or any application) can access a database, it needs to know which application created the database (Oracle, Excel, Paradox, etc.), and where the database file is located (its path on your system.) When you define the necessary information, AutoCAD can interface with external databases and thereby link records in tables to objects in drawings.

Once you define the information file that lists the database's application and location, you can then access its data from AutoCAD. This is true even if you do not have the database application installed on your PC. This means, for example, that you can access data in an Oracle database even though you don't have Oracle installed on your system. All you need is the database file that was created in Oracle. AutoCAD 2002 is designed to access external database files without using the originating database application.

AutoCAD 2002 is designed to access data from the following database applications:

- Microsoft Access 97
- dBase V and III
- Microsoft Excel 97
- Oracle 8.0 and 7.3
- Paradox 7.0
- Microsoft Visual FoxPro 6.0
- SQL Server 7.0 and 6.5

If you are provided database files created by any of these database applications, you can access the data in the files directly from AutoCAD 2002.

 Note: Microsoft Excel is not a true database application and therefore it contains no tables. Consequently, to access Excel data from within AutoCAD, you must first specify at least one named range of cells to act as a database table.

A single Excel file can contain multiple ranges of cells, with AutoCAD treating each range as a unique table.

To define the information AutoCAD 2002 needs to access external database information, you must do two things. First, you must use Microsoft's ODBC Data Source Administrator to create a *data source*, which lists the database's application type and its location. (The ODBC Data Source Administrator is accessed through the Windows Control Panel.) Second, you must use Microsoft's OLE DB application to establish a link between AutoCAD and the data source. (The OLE DB application is accessed from within AutoCAD though the Data Link Properties dialog box.) By using these two features, you can define the information AutoCAD needs to access external databases successfully.

Note: ODBC stands for Open Database Connectivity, which is a standard protocol for accessing information in SQL (Structured Query Language) database servers such as the Microsoft SQL Server.

OLE DB technology is newer than ODBC and performs a link-to-database function similar to that of ODBC. OLE DB, however, is designed to address issues encountered when working with non-relational database files, or when accessing distributed databases across the Internet, Intranets, and Extranets.

In the following two sections you will use the ODBC Data Source Administrator and the Data Link Properties dialog box to create the information necessary for AutoCAD to access data from an external database file.

CREATING AN ODBC DATA SOURCE FILE

To provide AutoCAD access to data in an external database file, you must first use the ODBC Data Source Administrator to create a data source file. The data source file identifies the database file's application type and the folder in which the database file is located. The data source file is a gateway that AutoCAD uses to access a database file. When you use the ODBC Data Source Administrator, you create a data source file that lets AutoCAD connect directly to a database file.

Note: If you work with Microsoft Access, Oracle, or Microsoft SQL Server database files, you can bypass setting up an ODBC data source and instead use the OLE DB direct drivers available on your system. Using the direct drivers you access the database files through an OLE DB (.udl) configuration file, which you create using AutoCAD's Data Link Properties dialog box.

For more information, refer to AutoCAD 2002's *Driver and Peripheral Guide*, in the section titled "Configure External Databases." Using this section's Procedures tab you can access the information necessary to set up a direct OLE DB configuration file. The guide is located in the acad_dpg.chm file, which is stored in the AutoCAD 2002/Help folder.

In the following example you will use the ODBC Data Source Administrator to create a data source for dBase files.

USING THE ODBC DATA SOURCE ADMINISTRATOR

1. Create a new folder on your PC called DB Files.

2. Copy the Manholes.dbf and Pipes.dbf files from the accompanying CD to the DB Files folder. After copying the dBase files, be sure to right-click on the *.dbf files, choose Properties, and then clear the Read-only attribute.

Next you will create a data source file using the two *.dbf files.

3. From the Windows Taskbar, choose Start>Settings>Control Panel. The Control Panel folder opens.

4. From the Control Panel, double-click on the Data Sources (ODBC) icon. The ODBC Data Source Administrator is displayed.

5. In the ODBC Data Source Administrator, choose the User DSN tab.

Note: An ODBC data source stores information on how AutoCAD connects to database files. There are three different methods for defining a data source:

- The User DSN folder creates a data source that is visible only to you and that can only be accessed from the computer on which the data source is created.

- The System DSN folder creates a data source that is visible to all users who have access rights to the computer on which the data source is created.

- The File DSN folder creates a data source that can be shared with other users who have the same ODBC drivers installed on their computer systems.

6. Choose Add. The Create New Data Source dialog box is displayed.

7. From the list of available database drivers, choose Microsoft dBase Driver (*.dbf), as shown in Figure 2–1.

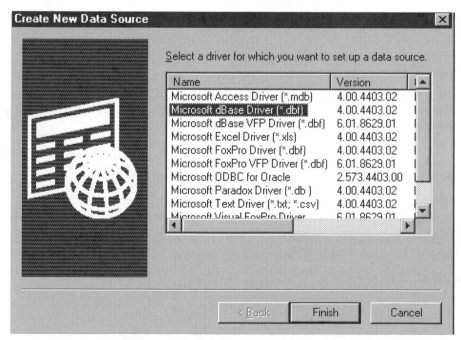

Figure 2–1 *The Create New Data Source dialog box identifies the driver to use to access database files.*

8.   Choose Finish. The ODBC dBase Setup dialog box is displayed.

By selecting the Microsoft dBase Driver, you indicated the database file's application type (dBase). Next you will identify the database file's location (its path.) Through the ODBC dBase Setup dialog box you locate the database file you wish to access and you also assign a name and description that makes it easy to identify the new data source from within AutoCAD.

9.   In the Data Source Name text box, type **StormDrains**.

10.   In the Description text box, type **Storm Drain Tables**.

11.   In the Database area, choose dBase 5.0 from the Version list. This is the appropriate version to select because the database files you copied to the DB Files folder are dBase 5.0 files.

12.   In the Database area, clear the Use Current Directory check box. The Select Directory button is activated, as shown in Figure 2–2.

13.   In the Database area, choose the Select Directory button. The Select Directory dialog box is displayed.

14.   In the Select Directory dialog box, browse to your DB Files directory, as shown in Figure 2–3.

15.   Choose OK to dismiss the Select Directory dialog box, then choose OK to dismiss the ODBC dBase Setup dialog box. The ODBC Data Source Administrator displays the new StormDrains data source, as shown in Figure 2–4.

16.   Choose OK. The data source is saved and can now be accessed from AutoCAD.

17.   You may close the Control Panel folder.

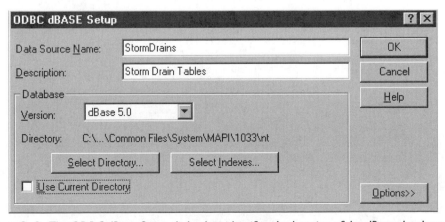

Figure 2–2 *The ODBC dBase Setup dialog box identifies the location of the dBase database files and assigns a data source name.*

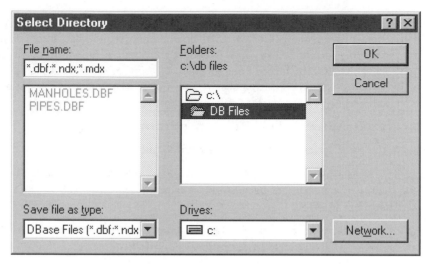

Figure 2–3 *The Select Directory dialog box allows you to locate the folder that contains the database files you use in AutoCAD.*

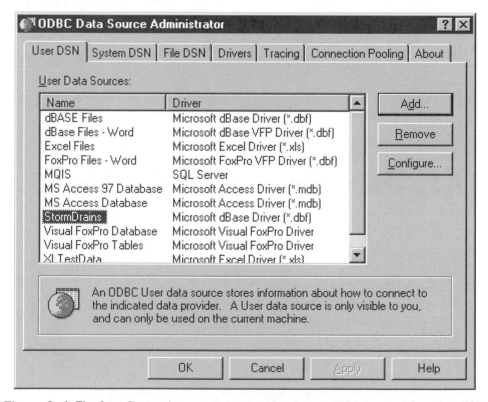

Figure 2–4 *The StormDrains data source is created and can now be accessed from AutoCAD.*

Using the ODBC Data Source Administrator you can create multiple data source files, with each data source file providing the information necessary for AutoCAD to access database files. When you define the data source files, you tell AutoCAD the type of database application that created the files and where the files are located on your PC. By using the ODBC Data Source Administrator to create a data source file, you can access data stored in database files from within AutoCAD.

Note: The steps used in the preceding example are those necessary to create data source files for accessing dBase files. The steps for accessing database files created in other applications vary slightly.

For more information, refer to AutoCAD 2002's *Driver and Peripheral Guide*, in the section titled, "Configure External Databases." The guide is located in the acad_dpg.chm file, which is stored in the AutoCAD 2002/Help folder.

In the next section you will use the data source file you just created to define an OLE DB configuration (.udl) file, which is the final component necessary for AutoCAD to access database files.

CREATING AN OLE DB CONFIGURATION FILE

An OLE DB configuration file contains information that AutoCAD uses to access data in database files. The configuration file is where you add information that identifies the ODBC data source file name, that lets you password-protect the connection, and that defines the location of the source database files. When you create an OLE DB configuration file, you give AutoCAD the information it needs to access database files.

In the following exercise you will create an OLE DB configuration file.

EXERCISE: CREATING AN OLE DB CONFIGURATION FILE

1. From the AutoCAD Tools menu, choose dbConnect. AutoCAD loads the dbConnect pull-down menu, inserting it into AutoCAD's menu bar, and then loads the dbConnect Manager window.

Note: The dbConnect command located on the Tools menu is a toggle. Once it is selected, AutoCAD places a check mark next to the dbConnect command. To remove the dbConnect menu and the dbConnect Manager, choose dbConnect from the Tools menu to clear the check mark.

2. From the dbConnect pull-down menu, choose Data Sources>Configure. AutoCAD displays the Configure a Data Source dialog box.

3. In the Data Source Name text box, enter **Storm_Drain**, as shown in Figure 2–5. This is the name you assign to the OLE DB (.udl) configuration file, and it is the name that will appear in the dbConnect Manager.

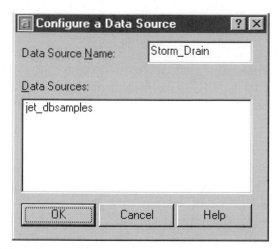

Figure 2–5 *AutoCAD adds the Data Source Name to the dbConnect Manager, which lets you connect to a database.*

4.   Choose OK. The Data Link Properties dialog box is displayed.

5.   In the Provider folder, select Microsoft OLE DB Provider for ODBC Drivers, as shown in Figure 2–6, and then choose Next. The Connection folder is displayed.

6.   In the Connection folder, under step 1, choose the Use Data Source Name radio button.

7.   From the Use Data Source Name list, choose the StormDrains data source name.

8.   Under step 2, select the Blank Password check box.

9.   Under step 3, choose the DB Files catalog from the list, as shown in Figure 2–7.

10. Choose the Test Connection button. AutoCAD performs a test to ensure that it is successfully connected to the database files. If it is, it displays a message noting that the test connection succeeded.

     If the test connection is successful, choose OK to dismiss the dialog box. If the text connection fails, you must check to ensure that you chose the correct data source name and the correct catalog folder.

11. Choose OK to dismiss the Data Link Properties dialog box.

     Once the Data Link Properties dialog box is dismissed, the Storm_Drain OLE DB configuration file appears in the dbConnect Manager, as shown in Figure 2–8. This means that you successfully configured the database and that AutoCAD can now access the two tables located in the DB Files folder.

12. You may close the drawing without saving your changes.

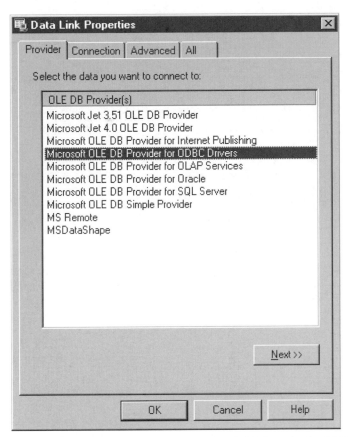

Figure 2–6 *In the Provider folder, select Microsoft OLE DB Provider for ODBC Drivers.*

Note: The StormDrains data source was created in the example located in the previous section, "Creating an ODBC Data Source File."

Tip: If the StormDrains data source name does not appear in the Use Data Source Name list, choose the Refresh button to update the list.

Once you successfully configure the database for use with AutoCAD, a configuration file with the extension .udl is created. This configuration file contains the information AutoCAD needs to access the configured database.

In the preceding two sections you created a data source for a dBase database and configured an OLE DB file to allow AutoCAD to access the dBase files stored in the DB Files folder. Next you will learn how to access the data in the database files.

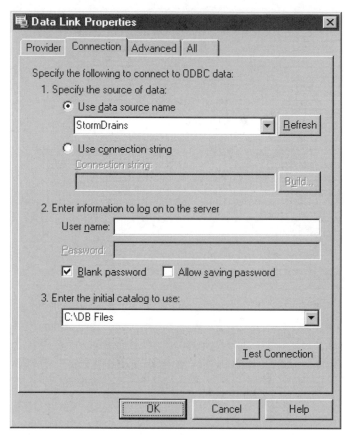

Figure 2–7 *The Connection folder lets you identify the data source name, password-protect access to the database files, and locate the folder, or catalog, that holds the database files.*

  Note: The DB Files catalog is the location (path) of the dBase files. The location was defined when the StormDrains data source was created in the example located in the previous section, "Creating an ODBC Data Source File."

  Note: A catalog in a database refers, in essence, to the folder that contains the database files.

WORKING WITH DATABASE TABLES

AutoCAD provides tools for working with external database files. By using these tools, you can view database tables and edit their data. When viewing tables, you can move, resize, and hide columns. You can specify a sort order for the table's records. You can edit or delete a record, and you can add a new record to a table. You can also search the table for particular values and then replace those values once they

Figure 2–8 *The newly created Storm_Drain OLE DB configuration file appears in the dbConnect Manager.*

are found. By using AutoCAD's database table tools, you can control the display of table data and edit the data in tables.

ACCESSING TABLES FROM THE DBCONNECT MANAGER

The dbConnect Manager provides a simple, visual way to review drawings linked to database files and to review the data sources to which the database files are linked. You can display links, labels, and queries defined in drawings. You can also display the tables associated with data sources. Through the dbConnect Manager you can easily view the tables available on your system and view the relationship between the tables and drawings to which they are linked.

 Tip: By default, AutoCAD stores the .udl files that it creates in the Data Links folder, which is located in the AutoCAD 2002 folder. You can specify a different location for .udl files from the Options dialog box.

 Note: To remove an OLE DB configuration file from AutoCAD's dbConnect Manager, you must delete its .udl file from the Data Links folder.

The dbConnect Manager lets you connect to data sources. Once connected, you can select a table in the data source and view or edit its data. You can also create a new link template or a new label template, which is used to link objects in drawings to

Note: There are some limitations with using AutoCAD's database connectivity feature, because AutoCAD uses a limited subset of the ODBC and OLE DB database configuration utilities.

For more information, refer to AutoCAD 2002's *Driver and Peripheral Guide*, in the section titled "Configure External Databases," under the heading "Known Limitations of Configuring External Databases." The guide is located in the acad_dpg.chm file, which is stored in the AutoCAD 2002/Help folder.

data in tables, and then label the objects using data extracted from the tables. You can also define a new query or execute an existing one. By using the dbConnect Manager, you can perform several useful functions on tables and their data.

To view the dbConnect Manager you can either select dbConnect from the Tools menu or type DBCONNECT at the command prompt. Once the command is executed, AutoCAD loads the dbConnect pull-down menu and the dbConnect Manager. The Manager is a dockable, resizable window that can float over your AutoCAD drawing windows. To close the Manager, either execute the dbConnect command again, or choose the Exit button (the small "X" button in the upper right-hand corner of the dbConnect Manager window). Once the Manager is closed, both it and the dbConnect pull-down menu are removed.

In the following exercise you will use the dbConnect Manager to connect to a data source and then view one of its tables.

EXERCISE: USING THE DBCONNECT MANAGER

1. From the AutoCAD Tools menu, choose dbConnect. AutoCAD loads the dbConnect pull-down menu and the dbConnect Manager.

 Note: The dbConnect menu and the dbConnect Manager are already loaded if you are continuing from the previous section's example.

2. Under Data Sources, right-click on jet_dbsamples, then choose Connect from the shortcut menu. AutoCAD connects to the jet_dbsamples data source and displays its tables, as shown in Figure 2–9.

    The jet_dbsamples data source is an OLE DB (.udl) configuration file that is automatically installed with AutoCAD 2002. The file is located in the AutoCAD 2002/Data Links folder.

3. Under the jet_dbsamples data source, choose the Employee table. The table is highlighted, and several tool buttons are activated on the dbConnect Manager, as shown in Figure 2–10.

4. Choose the View Table button. The Data View window is displayed in read-only mode, as shown in Figure 2–11.

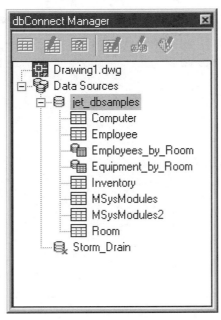

Figure 2–9 *When you connect to a data source, the dbConnect Manager displays its tables.*

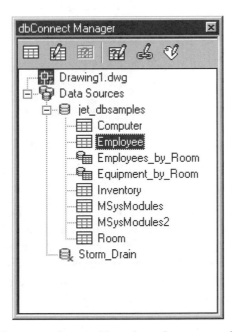

Figure 2–10 *When you select a table under a data source, the tool buttons are activated on the dbConnect Manager.*

	Emp_Id	Last_Name	First_Name	Gender	Title	Department	Room
▶	1000	Torbati	Yolanda	F	Programmer	Engineering	6044
	1001	Kleinn	Joel	M	Programmer	Engineering	6058
	1002	Ginsburg	Laura	F	President	Corporate	6050
	1003	Cox	Jennifer	F	Programmer	Engineering	6056
	1005	Ziada	Mauri	M	Product Designer	Engineering	6055
	1006	Keyser	Cara	F	Account Executive	Sales	6106
	1063	Ford	Janice	F	Accountant	Accounting	6020
	1010	Smith	Roxie	M	Programmer	Engineering	6054
	1011	Nelson	Robert	M	Programmer	Engineering	6042
	1012	Sachsen	Lars	M	Support Technician	Product Support	6104
	1013	Shannon	Don	M	Product Designer	Engineering	6053
	1016	Miro	Terri	F	Network Administrator	IS	6074
	1017	Lovett	Greg	M	Programmer	Engineering	6060
	1018	Larson	Steve	M	Programmer	Engineering	6052
	1019	Haque	Cintra	F	Sales Representative	Sales	6124
	1020	Sampson	Heather	F	Marketing Representative	Marketing	6076
	1023	Gupta	Rebecca	F	Programmer	Engineering	6051
	1024	Feaster	John	M	Vice President	Sales	6153

Data View - Employee (Drawing1.dwg) - Read Only

-- New Link Template --   -- New Label Template --

Record 1

Figure 2–11 *The Data View window may be accessed from the dbConnect Manager.*

Note: If you did not install AutoCAD in its default directory, you may need to update the jet_dbsamples.udl configuration file before working with these tables. Updating the file is only necessary if AutoCAD cannot locate the files on your system.

5.   Leave the Data View window open to continue with the next example.

The dbConnect Manager provides an easy way to display information about drawing links, labels, and queries, and information about data sources available on your system. From this window you can access table data and define queries. Using the dbConnect Manager you can quickly perform common database tasks.

In the next section you will use the Data View window to view data in a table.

VIEWING TABLE DATA WITH DATA VIEW

AutoCAD's Data View window allows you to manipulate its view to display data in the table the way you want. For example, you can hide and unhide columns. You can move columns to new positions in the table, changing their order. You can also sort the data in columns, in either ascending or descending order. You can even control the appearance of the text displayed in the table by changing its font or font size. Through the Data View window, you can manipulate the display of a table to make viewing data easier.

Note: When you choose the View Table button, AutoCAD also displays the Data View menu, which appears in AutoCAD's menu bar.

In the following exercise you will learn about controlling the appearance of data in the Data View window.

EXERCISE: MANIPULATING THE DATA VIEW WINDOW

1. Continue from the previous example and choose the Title column's header. You may need to scroll to the right in order to see the Title column.

   The column header is actually a button at the top of a column that displays the column's name. When you choose the button, AutoCAD highlights the entire column.

2. Drag the column to the left side of the Gender column. When a red line appears between the First Name and Gender columns, release your mouse button to drop the column. AutoCAD places the Title column to the left of the Gender column, as shown in Figure 2–12.

   You can resize a column as desired. By clicking and dragging the line between header buttons, you can resize a column. You can also automatically resize a column by double-clicking the line between two columns.

   AutoCAD lets you hide columns in the Data View, and you can unhide hidden columns. Next you will hide and then unhide the First_Name column.

3. Choose the First_Name column's header. The column is highlighted.

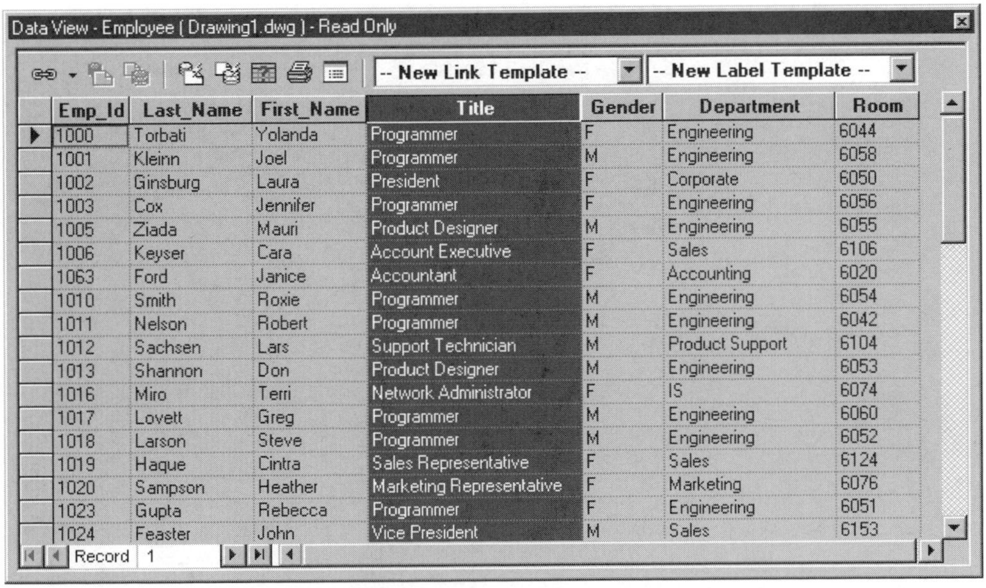

Figure 2–12 *You can highlight and move a column by picking and dragging its header, which is the button that displays the column's name.*

4.   Right-click over the First_Name column's header, then choose Hide from the shortcut menu. The First_Name column is hidden, as shown in Figure 2–13.

 Tip: By holding down the Ctrl key when choosing column headers, you can select and then hide multiple columns at the same time.

5.   Right-click over the Emp_ID column, then choose Unhide All from the shortcut menu. The First_Name column is redisplayed.

The Data View window allows you to sort the data in columns. The Data View window can sort both numerically and alphabetically, and it allows you to sort data in ascending or descending order. You can also execute sorts based on up to five columns at one time. Next you will sort two different columns at once.

6.   Right-click over the Emp_ID column's header button, then choose Sort from the shortcut menu. The Sort dialog box is displayed.

7.   In the Sort By area, choose Emp_ID from the list. Make sure the Ascending radio button is selected.

8.   In the Then By area immediately below the Sort By area, choose Department from the list, as shown in Figure 2–14. Make sure the Ascending radio button is selected.

Emp_Id	Last_Name	Title	Gender	Department	Room
1000	Torbati	Programmer	F	Engineering	6044
1001	Kleinn	Programmer	M	Engineering	6058
1002	Ginsburg	President	F	Corporate	6050
1003	Cox	Programmer	F	Engineering	6056
1005	Ziada	Product Designer	M	Engineering	6055
1006	Keyser	Account Executive	F	Sales	6106
1063	Ford	Accountant	F	Accounting	6020
1010	Smith	Programmer	M	Engineering	6054
1011	Nelson	Programmer	M	Engineering	6042
1012	Sachsen	Support Technician	M	Product Support	6104
1013	Shannon	Product Designer	M	Engineering	6053
1016	Miro	Network Administrator	F	IS	6074
1017	Lovett	Programmer	M	Engineering	6060
1018	Larson	Programmer	M	Engineering	6052
1019	Haque	Sales Representative	F	Sales	6124
1020	Sampson	Marketing Representative	F	Marketing	6076
1023	Gupta	Programmer	F	Engineering	6051
1024	Feaster	Vice President	M	Sales	6153

Data View - Employee (Drawing1.dwg) - Read Only

-- New Link Template --   -- New Label Template --

Record 1

Figure 2–13 *The First_Name column is hidden from view.*

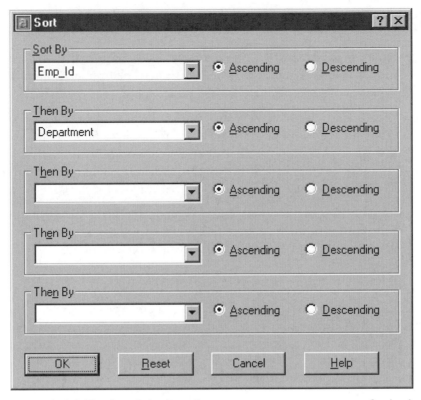

Figure 2–14 *The Sort dialog box allows you to create a sort on up to five levels.*

9. Choose OK. AutoCAD executes the sort and redisplays the data based on the sort results.

10. Close the Data View window and the dbConnect Manager by choosing the Exit button, which is the small button with an "X" in it, located in the upper-right corner of each window.

11. You may now exit AutoCAD without saving your changes.

As you work with large database files, some tables you display in the Data View window will contain many columns. To view these columns you will need to drag the slide bar at the bottom of the window left or right. The downside to dragging the slide bar to view columns outside the window's current display is that columns you are currently viewing will be moved out of view as you move the slide bar.

The Data View window allows you to freeze a column's position. This means that when you move the slide bar to view other columns, the frozen columns do not move from view, and their positions remain fixed. This feature allows you to drag the slide bar to view columns outside the current display, while always displaying the frozen columns.

To freeze a column, right-click over the column's header button, then choose Freeze from the shortcut menu. To unfreeze a column, right-click over any column's header button, then choose Unfreeze All.

The Data View window also allows you to control the alignment of text in columns and change the font and font size of the text used to display a table's data. To change the alignment of a column, right-click over the column's header button, choose Align from the shortcut menu, and then choose the desired alignment from the fly-out menu, as shown in Figure 2–15.

To control the font properties of text that appears in the Data View window, from the Data View pull-down menu, choose Format to display the Format dialog box, as shown in Figure 2–16. From the Format dialog box, you can modify font properties such as style and size.

The Data View window allows you to control its appearance. You can hide columns, change their position, or lock columns by freezing them. You can also sort the table's data using the Sort dialog box. In addition to these features, you can control the appearance of the text displayed in the columns and rows. You can change their justification and their font size. By manipulating its display, you can modify the Data View window to display the data the way you want.

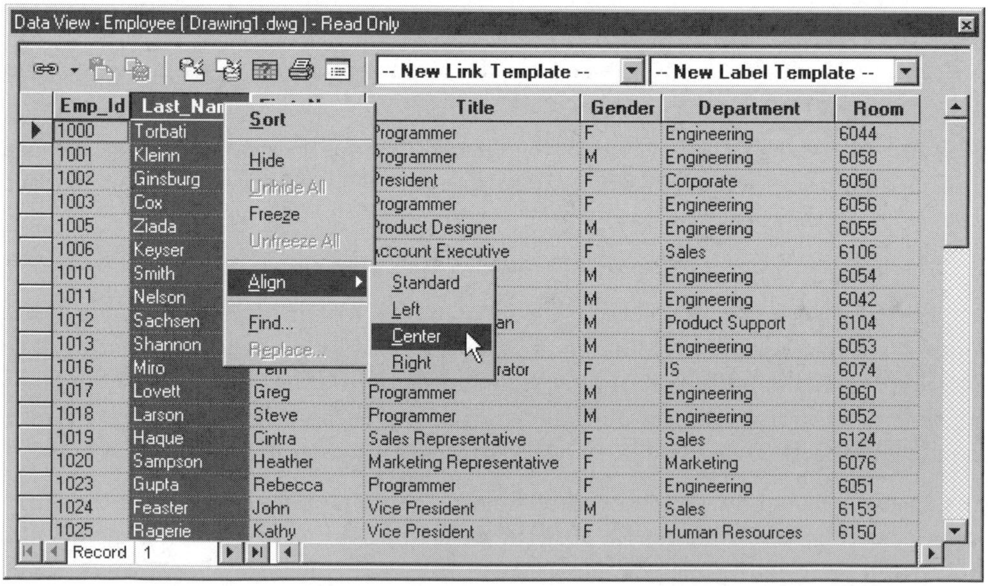

Figure 2–15 *Right-click over a column's header to control its text justification.*

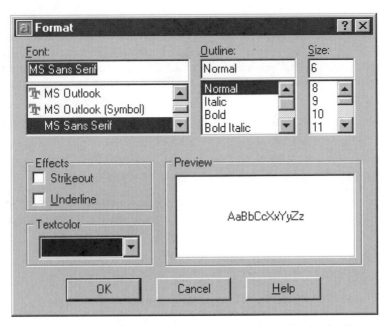

Figure 2–16 *The Format dialog box allows you to change the Data View window's font properties.*

EDITING TABLE DATA WITH DATA VIEW

AutoCAD's Data View window has two modes. The first is read-only mode, and the second is edit mode. The read-only mode allows you to view data and manipulate how it is displayed in the Data View window. The edit mode also allows you to manipulate the window's appearance but, more importantly, it allows you to manipulate the table and its data. For example, in edit mode, you can edit, delete, and add new records to a table. You can also search the table for certain values and replace those values with new ones. By using Data View's edit mode, you can view, edit, and explore the data in database tables.

In the following exercise you will learn about the various features of the Data View window's edit mode.

EXERCISE: USING DATA VIEW'S EDIT MODE

1. If you haven't done so in a previous example, create on your PC a new folder called DB Files.

2. Copy the StormDrain.mdb file from the accompanying CD to the DB Files folder. Then right-click on the *.mdb file, choose Properties, and clear the Read-only attribute.

The StormDrain.mdb file is a Microsoft Access 2000 database. As was noted previously in the section titled "Creating an ODBC Data Source File," it is not necessary to create an ODBC data source file to connect to Microsoft Access databases. Instead, you can use the OLE DB direct drivers available on your system to create an Access OLE DB (.udl) configuration file from within AutoCAD.

Next you will create an Access OLE DB (.udl) configuration file for the StormDrain.mdb database.

3. From the dbConnect pull-down menu, choose Data Sources>Configure. AutoCAD displays the Configure a Data Source dialog box.

4. In the Data Source Name text box, enter **StormDrains**. This is the name you assign to the Access OLE DB (.udl) configuration file, and it will appear in the dbConnect Manager.

5. Choose OK. The Data Link Properties dialog box is displayed.

6. In the Provider folder, select Microsoft Jet 4.0 OLE DB Provider, and then choose Next. The Connection folder is displayed.

7. In the Connection folder, under step 1, choose the ellipses button and then browse to the DB Files folder and open the StormDrain.mdb file.

8. Under step 2, select the Blank Password check box, as shown in Figure 2–17.

9. Choose the Test Connection button to confirm that AutoCAD is connected to the database files.

10. Choose OK to dismiss the Data Link Properties dialog box.

Once the Data Link Properties dialog box is dismissed, the StormDrains OLE DB configuration file appears in the dbConnect Manager.

11. Right-click on the StormDrains data source and then choose Connect. AutoCAD connects to the data source and displays its tables.

12. Choose the Pipes table and then choose the Edit Table button. AutoCAD opens the Data View window in edit mode.

From the Data View window you can edit individual cells in the table. Next you will edit the DIA value for one of the records.

13. In the DIA field for record 1008, replace value 24 with **36**, then press Enter. AutoCAD replaces the old cell value with the new one, as shown in Figure 2–18.

The Data View window lets you add new records to an existing table. Next you will add a new record.

14. Select the header for record 1010. (The record header is the small button to the left of the ID column.) AutoCAD highlights the record.

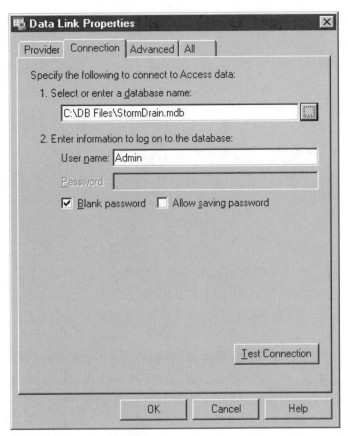

Figure 2–17 *The Connection folder lets you identify the Microsoft Access database file.*

Note: If the test connection is successful, choose OK to dismiss the dialog box. If the text connection fails, you must check to ensure that you chose the correct Microsoft Access database.

15.   Right-click on the highlighted record header, then choose Add New Record. AutoCAD adds a new, blank record at the end of the records list.

16.   Enter the following values for each cell in the new record:

ID: Type **1017**, then press Tab.

DIA: Type **24**, then press Tab.

TYPE: Type **RCP**, then press Tab.

D_LOAD: Type **1450**, then press Enter.

The Data View window should appear as shown in Figure 2–19.

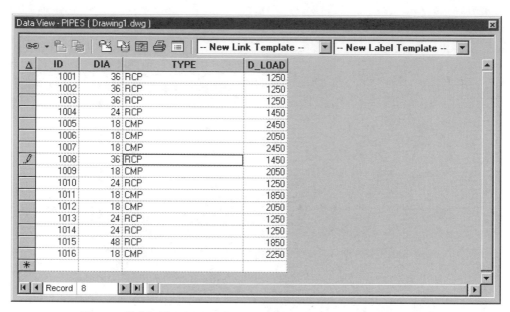

Figure 2–18 *The DIA field value for record 1008 is changed to 36.*

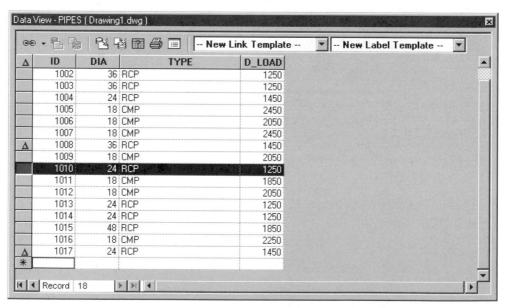

Figure 2–19 *The new record 1017 is added to the existing table.*

 Note: Once you enter a new value, AutoCAD instantly updates the table. Edits cannot be undone.

The Data View window lets you delete records from an existing table. Next you will delete a record.

17. Select the record header 1007. AutoCAD highlights the record.

18. Right-click on the highlighted record header, then choose Delete Record. AutoCAD displays a window asking you to verify that you want to delete the selected row.

19. Choose Yes. AutoCAD deletes the selected record.

 Tip: By holding down the Ctrl key when choosing record headers, you can select multiple records to delete at the same time.

The Data View window allows you to find specified values in selected columns. When you choose a cell in a column, right-click, and choose Find from the shortcut menu, AutoCAD will search the selected column and highlight the cell in which it finds the specified value. Additionally, you can replace the specified value with a new value.

Next you will search for values in the TYPE column and replace them with new values.

20. Select a cell in the TYPE column. AutoCAD highlights the cell.

21. Right-click over the cell, then choose Replace from the shortcut menu. AutoCAD displays the Replace dialog box.

22. In the Find What text box, type **CMP**.

23. In the Replace With text box, type **RCP**, as shown in Figure 2–20.

24. Choose Replace All. AutoCAD replaces all instances of the CMP value in the TYPE column with RCP.

25. Choose the Cancel button to dismiss the Replace dialog box. The Data View window displays the updated records, as shown in Figure 2–21.

26. Close the Data View window and the dbConnect Manager by choosing the Exit button, which is the small button with an "X" in it, located in the upper-right corner of each window.

27. You may now exit AutoCAD without saving your changes.

The Data View window's edit mode allows you to edit an existing table. You can change cell values individually, or you can search and replace specified values instantly in an entire column. You can add new records to the table or delete existing ones. By using the Data View window's edit mode, you can easily modify a table's data.

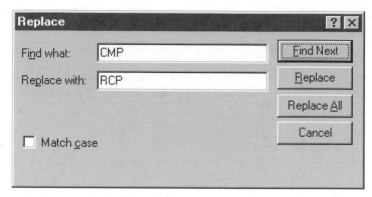

Figure 2–20 *The CMP value will be replaced by the RCP value.*

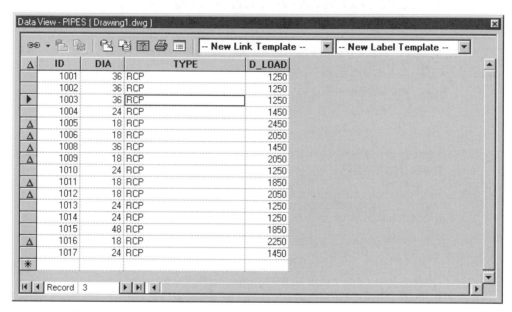

Figure 2–21 *The Data View window displays the table's updated records.*

 Note: AutoCAD only searches for values in the column indicated. AutoCAD cannot globally search all columns in a table for values.

WORKING WITH DATA AND OBJECTS

AutoCAD allows you to create a link between records in a database table and objects in a drawing. Once you have established a link between an object and a record, you can use the link to locate objects by selecting the records to which they are attached, or to locate records in a table by selecting objects in a drawing. Additionally, you can

Note: In this example, modifications made to the table instantly updated the original table. Other databases do not instantly update the original table; they allow you to choose whether you want to accept or reject your modifications before updating the original table.

When you work with databases that allow you to accept or reject your modifications, you are prompted to either commit modifications or restore the table's original values by right-clicking on the grid header and choosing the desired option from the shortcut menu. The grid header is the small button in the upper-left corner of the column and record headers.

extract the data values stored in a table and automatically insert them as a linked text label object. By using AutoCAD's ability to link data to objects, you can quickly locate linked objects and records and easily label objects by extracting the data to which they are attached.

LINKING DATA TO AUTOCAD OBJECTS

Up to this point you have learned how to use AutoCAD to connect to a database. By defining an ODBC data source file and then creating an OLE DB (.udl) configuration file, you can access a variety of databases. Through the data source and .udl files, you establish connections that let you view and edit databases from within AutoCAD. While the ability to connect to databases is useful, there is more to working with databases from within AutoCAD than just viewing and editing. The real power of connecting to databases from within AutoCAD lies in the ability to link objects in drawings to records in databases.

You may have noticed that in the previous examples in this chapter, while you connected AutoCAD to databases, the connections occurred independently from the current drawing. The fact is that while following the previous examples, you were simply connecting a database to the AutoCAD application but not to the AutoCAD drawing. Once you connect AutoCAD to a database, the next step is to link records in the database to objects in a drawing.

To link records in a database to objects in a drawing, you must first define a link template. Using a link template, AutoCAD lets you select a record in a table and link it to a graphic object in a drawing. So, by first connecting to a database through data source and .udl files, and then by defining a link template, you can create links between the graphic objects in a drawing and the records stored in database tables.

Creating a Link Template

A link template identifies the field (column) in a database's table to use to link a record to an object. By identifying the field in a table, you can then establish a link between a specific record's value and an AutoCAD object. For example, in the previous section you used AutoCAD to add a new record to an existing table in an Access database. The record included the unique ID value 1017. To link record 1017 to an object in a drawing, you must first create a link template that tells AutoCAD that the ID field contains the unique identifiers to use when linking a record to an object. So, after you connect a database to the AutoCAD application through the data source and .udl files, you then create a link template that tells AutoCAD which field's data to use to link table records in a database to objects in a drawing.

In the following exercise you will create a link template.

EXERCISE: CREATING A LINK TEMPLATE

1. Copy the P-02-F01.dwg file from the accompanying CD to the DB Files folder. The DB Files folder was created in a previous example in the section titled "Creating an ODBC Data Source File."

    After copying the drawing file, be sure to right-click on the file, choose Properties, and then clear the Read-only attribute.

2. Open the drawing P-02-F01.dwg located in the DB Files folder. The drawing contains a series of polylines and circles, representing storm drain lines and manholes.

3. From the Tools menu, choose dbConnect. AutoCAD displays the dbConnect pull-down menu and the dbConnect Manager.

4. In the dbConnect Manager, right-click on the StormDrains data source, and then choose CONNECT. AutoCAD connects to the data source and displays its tables.

5. From the dbConnect pull-down menu, choose Templates>New Link Template. AutoCAD displays the Select Data Object dialog box.

6. In the Select Data Object dialog box, choose the PIPES table as shown in Figure 2–22, then choose Continue. AutoCAD displays the New Link Template dialog box.

7. In the New Link Template dialog box, in the New Link Template Name text box, type PipesLink as shown in Figure 2–23, and then choose Continue. AutoCAD displays the Link Template dialog box.

8. In the Link Template dialog box, in the Key Fields list, choose the check box next to the ID key field as shown in Figure 2–24, and then choose OK. AutoCAD defines the new link template for the P-02-F01.dwg drawing, as indicated by the PipesLink connection displayed in the dbConnect Manager shown in Figure 2–25.

9. Save your changes and leave the drawing open for use in the next example.

When you work with databases from within AutoCAD, creating link templates is a necessary step that lets AutoCAD link records in a database to objects in drawings.

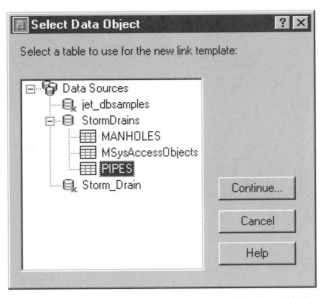

Figure 2–22 *The Select Data Object dialog box identifies the table for which you will create a link template.*

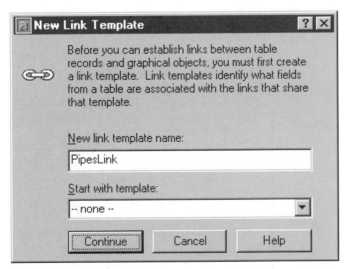

Figure 2–23 *The New Link Template dialog box lets you define the name for the new link template.*

When you create link templates that get stored within drawings, you identify the field in a table that AutoCAD uses to link records to objects.

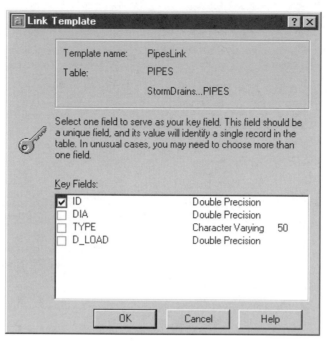

Figure 2–24 *The Link Template dialog box let you identify the key field for the new link template.*

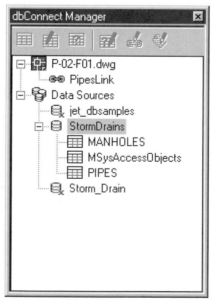

Figure 2–25 *The PipesLink link template is defined for the P-02-F01.dwg drawing and displayed in the dbConnect Manager.*

Note: Notice in Figure 2–25 that the PipesLink link template appears as an object under the P-02-F01.dwg drawing. As this hierarchy implies, the PipesLink link template is saved within the P-02-F01.dwg drawing. In contrast, the data sources shown in Figure 2–25 are independent of the P-02-F01.dwg drawing.

Next you will use the link template you just created to link records in the PIPES table to polyline objects in a drawing.

Creating Links

Once you define a link template for a drawing, you can then link records in a table to objects in a drawing. You can link multiple records to a single object, or you can link a single record to multiple objects. Once a link is created, you can select an object to highlight the record to which it is attached, or select a record to highlight the object to which it is linked. Once a link template is defined for a drawing, you can link records to objects and then view their link associations.

In the following exercise you will create links between records in a table and objects in a drawing.

EXERCISE: LINKING RECORDS TO OBJECTS

1. Continue from the previous example.

2. In the dbConnect Manager, choose the PIPES table, then choose the View Table button. AutoCAD displays the Data View window in read-only mode.

3. In the Data View window, from the New Link Template list, choose the PipesLink template.

4. In the Data View window, choose the record header for record ID 1001. AutoCAD highlights the record.

5. From the Data View pull-down menu (located in the AutoCAD menu bar), if it is not already checked, choose Link and Label Settings>Create Links. AutoCAD sets the link mode to create links.

6. From the Data View menu, choose Link!. AutoCAD prompts you to select objects to which it should link the highlighted record.

7. Choose the polyline that connects the circle labeled 1 and the circle labeled 3, then press Enter to end object selection. AutoCAD links the highlighted record to the selected polyline and changes the highlighted record's color to yellow.

8. In the Data View window, choose the record header for record ID 1004. AutoCAD highlights the record.

9. From the Data View menu, choose Link!. AutoCAD prompts you to select objects to which it should link the highlighted record.

10. Choose the polyline that connects the circle labeled 2 and the circle labeled 4, then choose the polyline that connects the circle labeled 4 and the circle labeled 5, and then press Enter to end object selection. AutoCAD links the highlighted record to the selected polylines and changes the highlighted record's color to yellow, as shown in Figure 2–26.

Once links are created between records and objects, you can view the links by either selecting records to highlight the objects to which they are attached, or by selecting objects to highlight the records to which they are linked.

Next you will use the link associations you just created to highlight records and objects.

11. In the Data View window, choose the record header for record ID 1001. AutoCAD highlights the record.

12. From the Data View menu, choose View Linked Objects. AutoCAD highlights the object to which the record is linked, as shown in Figure 2–27.

Next you will use an object to highlight a record.

13. Press Esc to deselect the highlighted polyline object.

14. Select the polyline that connects the circles labeled 4 and 5. AutoCAD highlights the polyline.

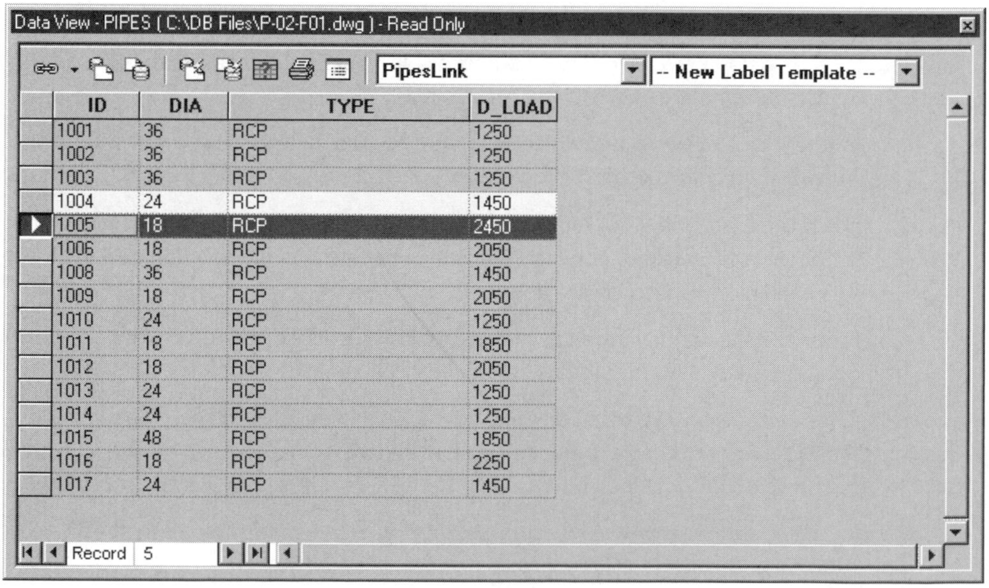

ID	DIA	TYPE	D_LOAD
1001	36	RCP	1250
1002	36	RCP	1250
1003	36	RCP	1250
1004	24	RCP	1450
1005	18	RCP	2450
1006	18	RCP	2050
1008	36	RCP	1450
1009	18	RCP	2050
1010	24	RCP	1250
1011	18	RCP	1850
1012	18	RCP	2050
1013	24	RCP	1250
1014	24	RCP	1250
1015	48	RCP	1850
1016	18	RCP	2250
1017	24	RCP	1450

Data View - PIPES (C:\DB Files\P-02-F01.dwg) - Read Only

PipesLink   -- New Label Template --

Record 5

Figure 2–26 In the Data View window, record 1004 is highlighted yellow after it is linked to an object in a drawing.

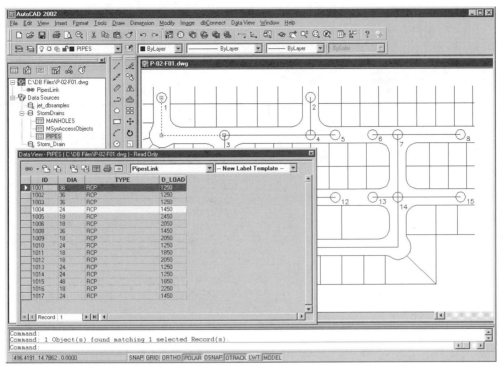

Figure 2–27 *The selected record is used to highlight the object to which it is linked.*

15. From the Data View menu, choose View Linked Records. AutoCAD displays only the record to which the object is linked in the Data View window, as shown in Figure 2–28.

16. Save your changes and then close the drawing.

Once you establish links between objects and records, you can use the links to locate and highlight linked objects. You can select objects and locate the records to which they are linked, or you can choose records to locate and select the objects to which they are attached. By creating links between records and objects, you add intelligence to your drawing that you can use to locate and select records and objects.

LABELING OBJECTS WITH DATA FROM TABLES

Once you have established links between records in a table and objects in a drawing, you can use those links to label the objects. AutoCAD provides a labeling feature that lets you use data in a table to place text in a drawing. By using AutoCAD's label feature, you can extract data from a table and use it to label the objects to which the data is linked.

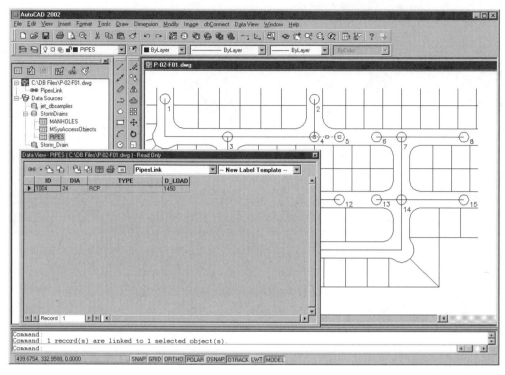

Figure 2–28 *The selected object is used to locate and display the record to which it is linked.*

Note: AutoCAD allows you to create freestanding labels, which are labels (text) that are inserted in a drawing using data from tables but are not attached to an object in a drawing.

Creating labels from data linked to objects requires two steps. First, you must create a label template. Second, you must insert the label. By performing these two simple steps, you can insert a label into a drawing, and attach it to an object in a drawing.

Note: The following example uses a drawing modified in previous examples. To perform the steps in the following example, you must use the drawing modified in the two examples located in the section titled "Linking Data to AutoCAD Objects."

In the following exercise you will use the links created between records in a table and objects in a drawing to label the objects.

EXERCISE: LABELING OBJECTS LINKED TO RECORDS

1. Open the P-02-F01.dwg file located in the DB Files folder.

2. If the dbConnect Manager is not displayed, from the tools menu, choose dbConnect. AutoCAD displays the dbConnect Manager and the dbConnect pull-down menu.

3. From the dbConnect menu, choose Templates>New Label Template. AutoCAD displays the Select a Database Object dialog box.

 Note: A label template requires a link template. You can only create a label template after you have created a link template.

4. In the Select a Database Object dialog box, make sure the PipesLink link template is selected, as shown in Figure 2–29, and then choose Continue. AutoCAD displays the New Label Template dialog box.

5. In the New Label Template dialog box, in the New Label Template Name text box, type **PipesLabel**, as shown in Figure 2–30, and then choose Continue. AutoCAD displays the Label Template dialog box.

 Note: The Label Template dialog box is actually the Mtext dialog box, and the label is inserted as an Mtext object.

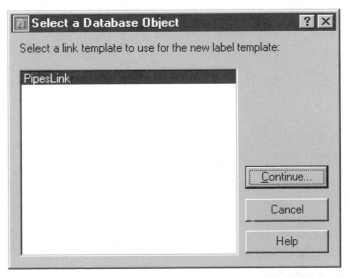

Figure 2–29 *The Select a Database Object dialog box is where you identify the link template to use with the label template.*

Figure 2–30 *The New Label Template dialog box lets you enter a name for the new label template.*

6.   In the Label Template dialog box, choose the Label Fields tab.

7.   From the Field list, choose DIA, then choose Add. AutoCAD adds the DIA field to the text window.

Note: The field value displayed in the text area represents the column from which AutoCAD will extract the value of the record that is linked to the selected AutoCAD object. You can add additional text, such as a prefix or suffix, to the field value displayed in the text window if desired.

8.   In the text window, after #(DIA), type a double quote symbol (").

9.   In the Label Template dialog box, choose the Properties tab.

10.   From the Justification list, choose Middle Left ML. The Mtext object's text insertion point is set to middle left.

11.   In the Label Template dialog box, choose the Character tab.

12.   Right-click in the text window, then choose Select All. The text is selected in the window.

13.   In the Font Height list, type **10.0**, as shown in Figure 2–31, then choose OK. AutoCAD defines the new label template for the P-02-F01.dwg drawing as indicated by the PipesLabel symbol displayed in the dbConnect Manager shown in Figure 2–32.

Next you will insert a label.

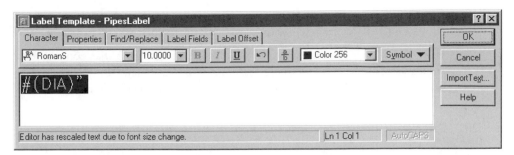

Figure 2–31 *The field from which table data is to be extracted is defined, and a quote symbol is added as a suffix to include with the label. Then, the highlighted text's height is set up to 10.0.*

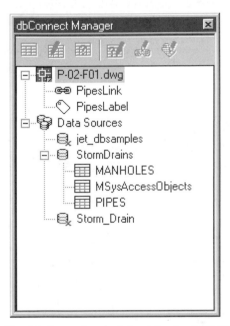

Figure 2–32 *The label template is defined for the P-02-F01.dwg drawing.*

14. In the dbConnect Manager, choose the PIPES table, then choose the View Table button. AutoCAD displays the Data View window in read-only mode.

15. From the New Link Template list, make sure the PipesLink link template is selected.

16. From the New Label Template list, make sure the PipesLabel label template is selected.

17. In the Data View window, select the record header for record ID 1003.

18. From the Data View pull-down menu, choose Link and Label Settings>Create Attached Labels. AutoCAD switches to "create attached labels" mode.

19. From the Data View menu, choose Link!, select the polyline that connects the circles labeled 3 and 4, and then press Enter. AutoCAD links the record to the selected polyline object and then inserts the label at the midpoint of the polyline, as shown in Figure 2–33.

20. Save your changes, then close the drawing.

AutoCAD creates a link between the selected record and the selected AutoCAD object, and then extracts the cell value from the field indicated by the label template. The label's insertion point is controlled in the Label Template dialog box from the Label Offset folder, and it may be modified. Additionally, you can edit label templates by choosing Templates>Edit Label Template from the dbConnect menu.

 Tip: After you insert the label, you can move the label above the polyline for easier viewing.

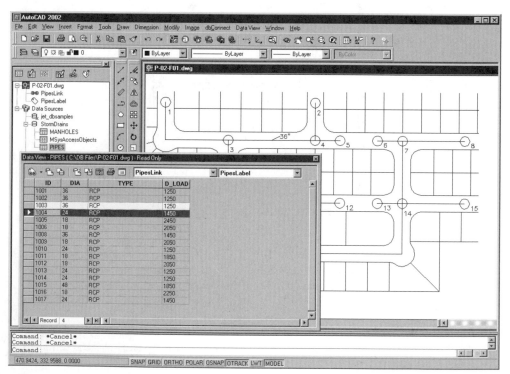

Figure 2–33 *The selected record is used to label the polyline.*

Tip: You can control the layer on which the label is inserted by switching to the desired layer before inserting the label.

By using AutoCAD's label feature, you can easily label objects in drawings by extracting values from data in tables. Using this technique you can automate the process of labeling objects and insert text values accurately by extracting the text values directly from the linked database table. AutoCAD's labeling feature makes adding text to drawings easier. In the next section you will use AutoCAD's query features to quickly find data.

USING QUERIES

Databases can contain enormous amounts of data. They can consist of dozens of tables, with each table containing hundreds of records, and each record consisting of many fields. The amount of data in a database can be overwhelming.

The goal of any database is to provide a place to organize and store large amounts of information, which you can then query for subsets of data. A query consists of search criteria that you specify. Once the search criterion is defined, the query is run, and the query searches the entire database for data that matches the specified criteria. If matching data is found, the query returns only those records that contain the matching data. By defining and running queries, you can quickly extract the particular set of data you need.

AutoCAD allows you to create and run queries. You can create queries that search through a table for specified values. You can also define queries that search through the objects in a drawing, returning a selection set of records that meet the query criteria.

You create queries in AutoCAD using the Query Editor, which consists of four folders, as shown in Figure 2–34. The folders are designed to build queries, and the purpose of each is described as follows:

- **Quick Query**—Lets you define and run a query using basic operators such as Is Equal To or Is Greater Than.

- **Range Query**—Lets you define and run a query based on a range of values. For example, you can query for all objects whose field value is greater than or equal to 18 and less than or equal to 36.

- **Query Builder**—Lets you define and run a query using multiple operators and ranges, and allows you to use parentheses to group the criteria. Additionally, you can use Boolean operators such as AND and OR to further refine your query. This folder represents AutoCAD's primary query builder.

- **SQL Query**—Lets you define and run a query by creating SQL statements that conform to Microsoft's implementation of the SQL 92 protocol. This folder allows you to build free-form SQL queries, queries that perform relational operations on multiple database tables using the SQL "join" operator.

By using AutoCAD's Query Editor, you can construct a variety of queries that range from the simple to the complex.

In the following section you will use the Query Editor to quickly define and run a simple query.

QUERYING OBJECTS

The simplest way to create a query is to use the Query Editor's Quick Query feature. The Quick Query feature allows you to define a basic query that can find data using simple comparison operators such as Is Equal To or Is Greater Than. Using the Quick Query feature, you can easily define and run a query that searches your data and returns the values you need.

 Note: The following example uses a drawing modified in previous examples. To perform the steps in the following example, you must use that drawing, which is modified in the examples located in the section titled "Working with Data and Objects."

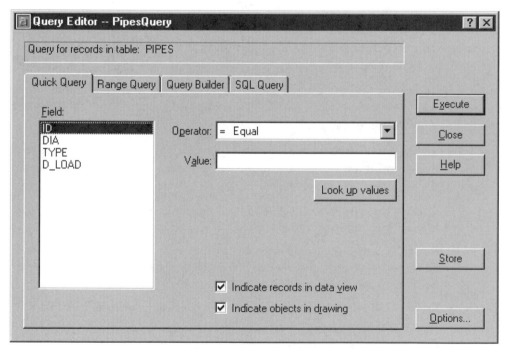

Figure 2–34 *The Query Editor lets you build queries that range from simple to complex.*

In the following exercise you will use the Quick Query feature to build and run a query that searches for pipes with a diameter equal to 36 inches.

EXERCISE: CREATING AND RUNNING A QUERY

1. Open the P-02-F01.dwg file located in the DB Files folder.

2. From the Tools menu, choose dbConnect. AutoCAD displays the dbConnect Manager and the dbConnect pull-down menu.

3. In the dbConnect Manager, right-click on the StormDrains data source, then choose Connect. AutoCAD connects to the data source and displays its tables.

4. From the dbConnect menu, choose Queries>New Query on an External Table. AutoCAD displays the Select Data Object dialog box.

5. In the Select Data Object dialog box, select the PIPES table as shown in Figure 2–35, then choose Continue. AutoCAD displays the New Query dialog box.

6. In the New Query dialog box, in the New Query Name text box, type PipesQuery as shown in Figure 2–36, and then choose Continue. AutoCAD displays the Query Editor.

7. In the Query Editor, select the Quick Query tab.

8. In the Quick Query folder, from the Field list, choose DIA.

9. From the Operator list, choose = Equal.

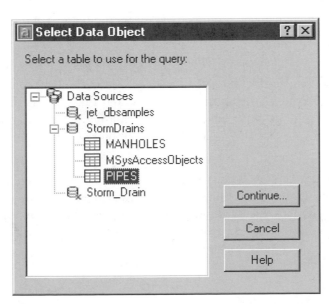

Figure 2–35 *From the Select Data Object dialog box you choose the table you wish to query.*

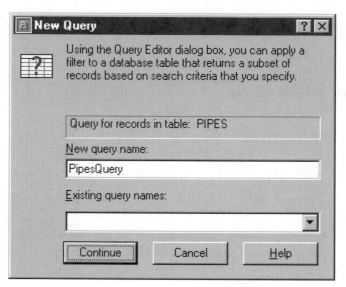

Figure 2–36 *From the New Query dialog box, you define the name for the new query.*

10. In the Value text box, type **36**.

11. Make sure the Indicate Records in Data View and the Indicate Objects in Drawing check boxes are selected, as shown in Figure 2–37.

12. Choose Execute. AutoCAD executes the query, returns the subset records that match the query criteria, and displays the records in the Data View window, as shown in Figure 2–38.

13. You may exit AutoCAD without saving your changes.

With the query's selection set returned and displayed in the Data View window, you could highlight all the records in the Data View, and then choose the View Linked Objects in Drawing button to have AutoCAD highlight the selection set's linked objects in the drawing, thereby visually locating all pipes with a diameter equal to 36 inches.

AutoCAD's Query Editor allows you to define and run a query. You can quickly define a query using the Quick Query folder, or you can create more complex queries using the Range Query, Query Builder, or SQL Query folder. By using the Query Editor, you can easily define and execute a query that returns the data you need.

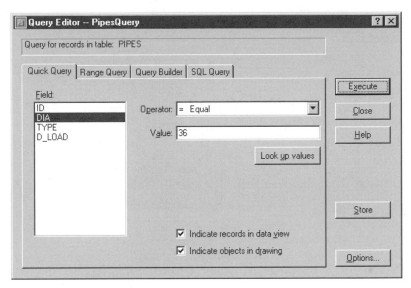

Figure 2–37 *The Quick Query is defined and ready to run.*

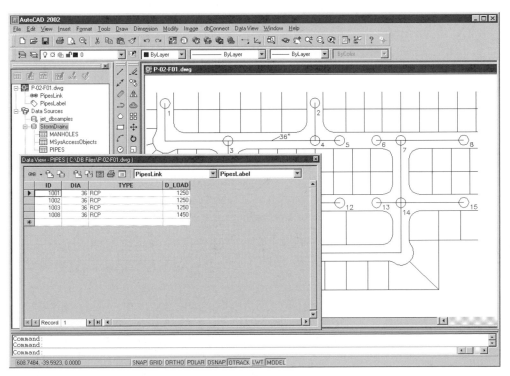

Figure 2–38 *The Quick Query returns the subset of records that match the query's criteria.*

SUMMARY

In this chapter you reviewed how to use the ODBC Data Source Administrator provided by the Windows operating system to create a data source for a database file. You learned how to use the Data Link Properties dialog box to create an OLE DB (.udl) configuration file, which lets AutoCAD connect to an external database. You used the dbConnect Manager to access external database tables, and you used the Data View window to view and edit the information in tables. You learned how to link data to AutoCAD objects and how to label objects with data extracted from external database tables. Finally, you learned how to execute queries and query for objects either in AutoCAD drawings or in tables.

By using the techniques discussed in this chapter, you can associate external database files with your AutoCAD drawings and use the tools provided by AutoCAD to work with the data stored in database tables.

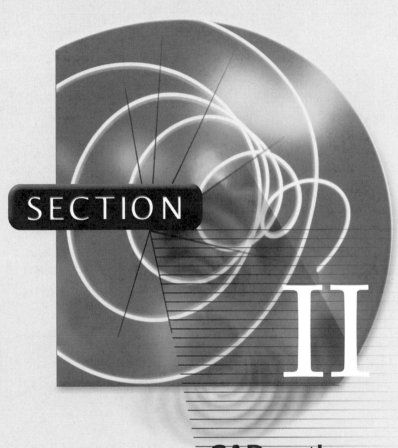

SECTION

II

CAD on the
Internet

CHAPTER 3

Transmitting Drawings and Publishing to the Web

AutoCAD 2000i emphasized integration with the Internet, hence the "i" tacked onto the end of "2000." In truth, however, Autodesk has been adding Internet features to releases of AutoCAD since DWF export and hyperlinking were added to Release 14 mid-release.

Sometimes you need to share your drawings with others. AutoCAD 2002 provides two tools for doing this conveniently. The eTransmit command packages files together for transmittal by email, while the PublishToWeb command creates a Web page for displaying drawings.

For example, your firm may want to advertise its drafting and design abilities. Posting examples of your drawings to a Web site allows others to see your work. Or, you may want to share drawings and symbol libraries with others in your office (using a local Intranet) and your clients (via an Extranet). The PublishToWeb command allows you to do this.

After completing this chapter you will be able to

- Collect drawings and support files into a package for transmittal
- Publish a drawing as a Web page

TRANSMITTING DRAWINGS

Emailing a drawing is not quite as easy as emailing, say, a Word document. That is because an AutoCAD drawing is not an island; it is usually interconnected to a number of other files. These might include fonts, font mapping, raster images, and externally referenced drawings. One function of the eTransmit command is to check for additional files that may need to accompany the drawing file(s).

The eTransmit command performs the following tasks:

- Finds all the files associated with the drawing
- Collects the files in a single compressed file, a self-extracting file, or a folder of files; allows you to lock the collection with a password
- Strips paths from xrefs and image files (optional)
- Provides an area for you to enter notes for the recipient
- Saves the drawing in AutoCAD2000 or Release 14 format
- Generates a Web page with a link to the files, instead of sending the package as email
- Produces a report that includes instructions to the recipient

(The eTransmit command, introduced with AutoCAD 2000i, is an expanded version of the Pack'n Go command, a "bonus" command provided with AutoCAD Release 14.)

TRANSMITTAL CAUTIONS

Including TrueType fonts (TTF) is a touchy issue, because doing so can infringe on copyrights. The online Help included with AutoCAD states that TTF fonts are not included in the transmittal, but in practice they are.

While drawings can be emailed, they can be very large, which means that they can take a long time to send or receive over a slow Internet connection, such as via a modem.

When you select eTransmit, the command prompts you to first save the drawing, as shown in Figure 3–1.

Figure 3–1 *Save Changes dialog box.*

EXERCISE

In this exercise you will learn how to create an electronic transmittal of an AutoCAD drawing. Start AutoCAD and open the office.dwg file found on the CD-ROM.

1. To create a transmittal, start the eTransmit command. From the menu, select File>eTransmit. Notice the Create Transmittal dialog box (Figure 3–2).

2. In the Notes section, type a message such as the following to your recipient:

   ```
   Here are the first-floor drawings with the revisions you
   requested. Call me at 555-1212 if you have any questions.
   — Andy Cadmann
   ```

3. Select the remaining options in the dialog box as follows:

   Type: Self-extracting executable

   Password: (Leave this blank.)

   Location: C:\

   Convert Drawing to: AutoCAD 2000

   Preserve Directory Structure: No

   Remove Paths from xrefs and Images: Yes

   Send Email with Transmittal: No

   Make Web Page Files: Yes

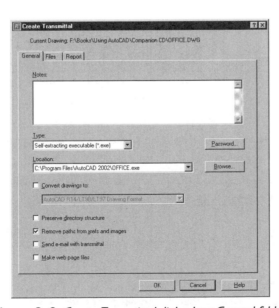

Figure 3–2 *Create Transmittal dialog box: General folder.*

The options are defined as follows:

- **Self-extracting executable**—Means that the files will be compressed into a single file with the extension .exe. The email recipient will double-click on the file name to extract (uncompress) the files. The benefit is that the recipient does not need to have a copy of PkUnzip or WinZip in their computer. The drawback is that a virus could hide in the .exe file.

- **Password**—Locks the transmittal so that only people who know the password can extract the files. It provides a simple level of security.

- **Location**—Specifies where the transmittal file should be stored on your computer. The C:\ location is handy for finding it easily. You may, however, want a separate folder in which all transmittals are located.

- **Convert Drawing to**—Saves the drawing in the current (2000, 2000i, and 2002) release or in previous (Release 14) versions of AutoCAD. The benefit to saving in Release 14 format is that clients with older versions of AutoCAD and AutoCAD LT can read the file. The drawback is that objects specific to AutoCAD 2000/2000i/2002 might be erased or modified to a simpler format, as follows:

  Lineweights would no longer be displayed (but would be restored when the drawing was opened again in AutoCAD 2000).

  Database links and freestanding labels would be converted to AutoCAD Release 14 links and displayable attributes.

- **Preserve Directory Structure**—Means that files will be extracted to the same folders from which they were collected. This option should be turned on only if the recipients have the identical folder structure on their computers.

- **Remove Paths from xrefs and Images**—Means that the recipient does not have to worry about xrefs and images being located in a specific folder.

- **Send Email with Transmittal**—Launches Windows' built-in email software automatically. If you use an alternative email program, turn off this option.

- **Make Web Page Files**—Generates a Web page. This lets the recipient download the transmittal file via their Web browser. This is useful in two cases: (1) when the recipient does not have email access, such as when the recipient is out in the field; (2) when you want the transmittal to be made available on a more public basis. (Recall that the file can be locked via the password.) In this tutorial we will create the Web page to see what it looks like.

4. Choose the Files tab. Notice the tree list of associated files (Figure 3–3). AutoCAD determined that the drawing depends on several other files. Imagine

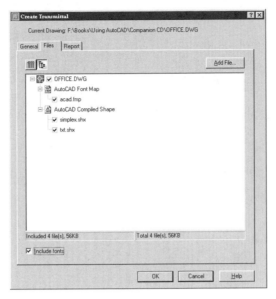

Figure 3–3 *Create Transmittal dialog box: Files folder.*

if you had to figure out that list on your own. Near the bottom of the dialog box you can view the summary:

```
Included 4 file(s), 56KB
```

The size given, 56KB, is the uncompressed total file size. Once the files are compressed, their size will be smaller.

5.   Click Include Fonts to turn *off* the option; the size becomes smaller:

```
Included 2 file(s), 29KB
```

The Include Fonts option determines whether AutoCAD includes the font files with the transmittal. Normally the option should be turned off for two reasons: (1) the fonts you use are copyrighted and cannot legally be copied; or (2) the recipient already has the fonts. Not including the fonts can also save a significant amount of file space. Recall that a smaller file takes less time to transmit.

6.   Choose the Report tab. Notice that AutoCAD has automatically generated a report that describes the files being sent, how to deal with external files (xrefs and so on), and other notes (Figure 3–4). Scroll through the report to familiarize yourself with its contents.

Choose Save As to save the report as a text file for yourself; the report will be included in the transmittal package.

7.   Choose OK. AutoCAD spends a few seconds putting together the transmittal package, as well as the Web page.

8.   You have three choices for sending the transmittal:

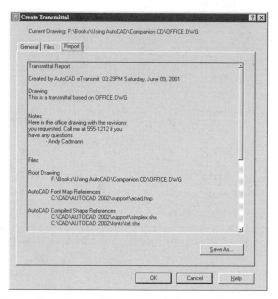

Figure 3–4 *Create Transmittal dialog box: Report folder.*

- **Email**—Emails the office.exe file to your recipient.

- **Web**—Uploads the files to a Web site via FTP (File Transfer Protocol). Your recipients use their Web browser to access the files.

- **Courier**—Leaves the office.exe as a file, which you copy to diskette and send by courier to your recipient.

 Note: Files occasionally become corrupted when they are sent by email. Ask the person who sent you the e-transmittal to resend it. If you continue to have problems, you may need to change a setting in your email software. For example, try changing the attachment encoding method from BinHex or Uuencode to MIME.

The Web page generated by AutoCAD is optimized for use with Internet Explorer. I found that the Web page also works with Netscape, Mozilla, and Groove, but not with Opera.

EXERCISE: RECEIVING THE E-TRANSMITTAL
In this exercise you will learn how to "read" the transmittal.

1. To open the transmittal, double-click on the office.exe file. Notice that a dialog box asks where to extract the files (Figure 3–5). Use a temporary folder, such as C:\Temp or C:\Windows\Temp and then choose OK.

2. Use Windows Explorer to view the contents of the Temp folder. Double-click on the office.txt report file to read it, as shown in Figure 3–6.

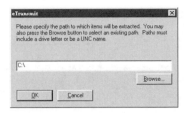

Figure 3–5 *eTransmit asks where to extract the files.*

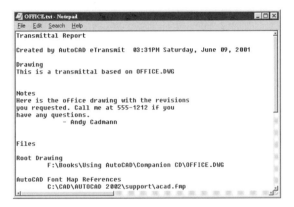

Figure 3–6 *The office.txt report describes the files sent by eTransmit.*

3.   Use your Web browser to view the page generated by the eTransmit command. Double-click on the office.htm file (Figure 3–7). Notice the links you can click on:

   •   **Download 1st Floor.exe**—Downloads the compressed transmittal file to the folder you specify. Follow steps 1 and 2 to uncompress the files.

   •   **View Report for 1st Floor**—Scrolls the Web page to show the report.

PUBLISHING DRAWINGS TO THE WEB

AutoCAD's Publish to Web feature is a "wizard" that helps you through the steps required to output a drawing as a Web page. In addition to using the Publish to Web wizard, you can "manually" export AutoCAD drawings using the DwfOut command, which simply exports the drawing to a DWF (short for "drawing Web format") file. You then have to add the HTML code that makes the DWF file appear in a Web page. You will learn more about DWF in the next chapter.

A third alternative is to use the Plot command's ePlot option, which performs essentially the same function as the undocumented DwfOut command.

Before exporting, you may want to place hyperlinks in the drawing using the Hyperlinks command. See Chapter 4, AutoCAD on the Internet, for more information.

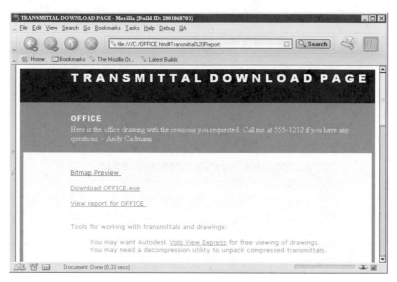

Figure 3–7 *Transmittal Report displayed by a Web browser.*

EXERCISE

In this exercise you will learn how to create a Web page from a drawing.

1. Start AutoCAD and then open the office.dwg file found on the CD-ROM.

2. Start the Publish to Web wizard with the PublishToWeb command. As an alternative, from the menu, select File>Publish to Web. Notice the Publish to Web wizard (Figure 3–8). The dialog box has two radio buttons (the round buttons near the center) used to select the type of wizard you will use:

   • **Create New Web Page**—Guides you through creating a new Web page.

   • **Edit Existing Web Page**—Guides you through editing a Web page previously created by this wizard.

3. Select Create New Web Page and then choose Next.

4. For this exercise, specify the following options (see Figure 3–9):

   Specify the Name of Your Web Page: **Using AutoCAD**

   Specify the Parent Directory: (Accept the default.)

   Provide a Description: **Example Web page created by AutoCAD**

   So that you can edit the Web page later, AutoCAD stores parameters in a file stored in the \Windows\Applications Data\Autodesk folder.

   On the Web page, the name will appear at the top, and the description will appear just below the name. Choose Next.

5. Select an image type, described as follows, from the drop-down list:

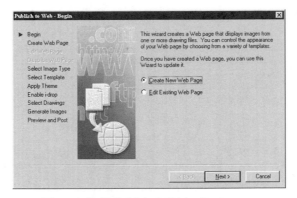

Figure 3–8 *Publish to Web - Begin page.*

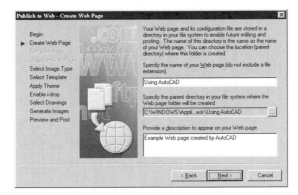

Figure 3–9 *Publish to Web - Create Web Page page.*

- DWF (drawing Web format)—A vector format that is displayed cleanly and can be zoomed and panned. Not all Web browsers can display DWF.

- JPEG (joint photographic experts group)—A raster format that all Web browsers display. It may create artifacts (details that do not exist).

- PNG (portable network graphics)—A raster format that does not suffer the artifact problem. Some older Web browsers do not display PNG.

The raster image sizes available for JPEG and PNG are shown in Table 3-1.

A larger image provides more detail but takes longer to transmit. For this exercise, select JPEG (Figure 3–10).

From the Image Size drop-down list, select Small. Then choose Next.

6. The Select Template page provides predesigned formats for the Web page (Figure 3–11). Select List Plus Summary and then choose Next.

Table 3–1: *Available Raster Image Sizes for JPEG and PNG*

Image Size	Resolution	Approximate PNG File Size
Small	789 x 610	60KB
Medium	1009 x 780	90KB
Large	1302 x 1006	130KB
Extra large	1576 x 1218	170KB

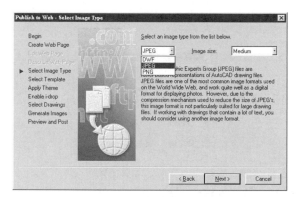

Figure 3–10 *Publish to Web - Select Image Type page.*

Figure 3–11 *Select a Web page template.*

7. Select a theme to apply to the Web page, such as "Autumn Fields" (see Figure 3–12). Choose Next.

8. As an option, you can enable Autodesk's "i-drop" feature in the Web page. For this exercise, leave the Enable i-drop box unchecked, and then choose Next.

9. Select the drawing(s) and specify related parameters, which are defined as follows (see Figure 3–13):

   • **Drawing**—Selects the drawing. If one or more drawings are open, their names are listed. You can also select drawings by choosing the ... button, which displays a file dialog box.

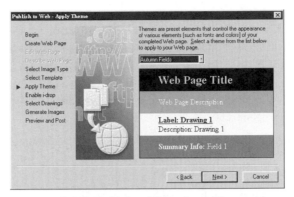

Figure 3–12 *Publish to Web - Apply Theme page.*

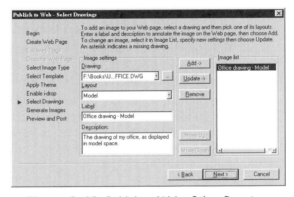

Figure 3–13 *Publish to Web - Select Drawings.*

- **Layout**—Selects a model or a layout mode. If the drawing contains one or more layouts, they are listed here.

- **Label**—Provides a name by which the drawing is known on the Web page. You could use the drawing's file name or another more descriptive name.

- **Description**—Provides a description that appears with the drawing on the Web page.

For this tutorial, enter the following:

Drawing: **office.dwg**

Layout: **Model**

Label: **Office drawing - Model**

Description: **The drawing of my office, as displayed in model space.**

Choose Add. Notice that the drawing is referred to by its label. If you change your mind, you can remove drawings from the list or move drawings up and down the list. When you are finished, choose Next.

10. To ensure that the drawings are up to date, AutoCAD regenerates them. Select the Regenerate All Images option, unless you have an exceptionally slow computer or a large number of drawings to process (Figure 3–14). Wait while AutoCAD regenerates the images. Choose Next.

11. Click Preview to see what the Web page will look like. AutoCAD launches your computer's default Web browser. If the page is not to your liking, choose the Back button and make changes.

12. Exit the Web browser. In AutoCAD, choose Finish to exit the wizard.

Note: When you choose the Preview button to launch the Web browser, you might not see the drawing, because the HTML code created by AutoCAD requires that JavaScript is turned on. Make sure your Web browser's JavaScript option is turned on.

If the image is a DWF file, the browser displays it only when Autodesk's Whip! plug-in has been added to the Web browser. Right-click on the image to see a menu that allows you to zoom and pan, toggle layers and named views, print, and so on.

The Post Now option works only if you have correctly set up the FTP (File Transfer Protocol) parameters. If you have, AutoCAD can directly upload the HTML files to your Web site. If you have not, use a separate FTP program to upload the files from your computer's \Windows\Applications Data\Autodesk folder or another folder defined in step 4 of this exercise.

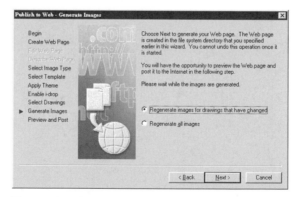

Figure 3–14 *Publish to Web - Generate Images page.*

EXERCISE: EDITING THE WEB PAGE

In the previous exercise you used the Publish to Web wizard to create a Web page from a drawing. In this exercise you will learn how to edit the Web page using the same wizard.

1. Start the PublishToWeb command, select Edit Existing Page, and then choose Next. Notice that the wizard displays a list of previously published Web pages. Choose either Browse or Preview, according to the following:

    • Choose Browse if the Web page created by AutoCAD is not on the list. AutoCAD displays the Select the PublishToWeb Project to Open file dialog box.

    • Choose Preview if you cannot remember what the Web page looks like. AutoCAD launches your computer's default browser and displays the selected Web page.

    Select the Web project you created in the previous exercise and then choose Next.

2. Change the title and the description of the Web page. If necessary, edit the wording. Then choose Next.

3. Change the format of the Web page. Choose Next.

4. Change the template and then choose Next.

5. Change the Web theme and then choose Next.

6. Toggle to select whether i-drop is enabled, and then choose Next.

7. You can change the descriptions of drawings and select different drawings. If you wish, make changes, and then choose Next.

8. Wait for AutoCAD to regenerate the drawings, then choose Preview to check the changes.

9. Exit the Web browser. In AutoCAD, choose Finish to exit the wizard.

 Note: If you cannot find the Web page generated by AutoCAD, look for it in the folder named \Windows\Applications Data\Autodesk, unless you specified a different folder.

The folder holds numerous HTML files generated by AutoCAD. The primary HTML file is called acwebpublish.htm. Drag this file into the Web browser to view the Web page and its drawing(s).

The Publish to Web wizard supports browsers that support JavaScript. I tested it successfully with Opera 5, Netscape 4, Netscape 6, and Internet Explorer 5.5. It might not work with browsers found on handheld devices, which tend not to have JavaScript support.

SUMMARY

In this chapter you learned how to use the eTransmit and PublishToWeb features. With eTransmit, you can quickly pack the current drawing's referenced files, including xrefs and fonts, into a single zip file that automatically includes a transmittal letter. Through the PublishToWeb feature, a wizard steps you through the process of posting your drawings on the web, where they can be shared over the Internet. Using these two powerful features, AutoCAD extends your project's reach to anyone, anywhere in the world.

CHAPTER 4

AutoCAD on the Internet

The Internet has become the world's most important way to exchange information. AutoCAD allows you to interact with the Internet in several ways. It can open and save files that are located on the Internet, launch a Web browser (AutoCAD 2002 includes a simple Web browser), and create DWF (drawing Web format) files for viewing as drawings on Web pages. After completing this chapter you will be able to

- Launch a Web browser from AutoCAD
- Understand the importance of URLs (uniform resource locators)
- Open and save drawings to and from the Internet
- Place hyperlinks in a drawing
- Convert drawings to DWF file format
- View DWF files with a Web browser
- Learn to use the WHIP! plug-in

INTRODUCTION

This chapter introduces you to the following Web-related commands:

Browser launches a Web browser from within AutoCAD.

Hyperlink attaches a URL to (and removes a URL from) an object or an area in the drawing.

> HyperlinkFwd moves you to the next hyperlink (undocumented command).
>
> HyperlinkBack moves you to the previous hyperlink (undocumented command).
>
> HyperlinkStop stops the hyperlink-access action (undocumented command).
>
> PasteAsHyperlink attaches a URL to an object in the drawing from text stored in the Clipboard (undocumented command).

You are probably already familiar with the most common uses of the Internet: sending email (electronic mail) and browsing the Web (short for "World Wide Web"). Email lets users exchange messages and data at a very low cost. The Web brings together text, graphics, audio, and movies in an easy-to-use format.

AutoCAD allows you to interact with the Internet in several ways, some of which you learned about in the previous chapter. In addition, AutoCAD can launch a Web browser from within AutoCAD via the Browser command. Hyperlinks can be inserted in drawings via the Hyperlink command, allowing you to link the drawing to other documents on your computer and the Internet.

AutoCAD can open, insert, and save drawings to and from the Internet via the Open, Insert, and SaveAs commands. With the Plot command's ePlot option (short for "electronic plot"), AutoCAD can create DWF files for viewing drawings in 2D format on Web pages.

UNDERSTANDING THE URL

The URL (short for "uniform resource locator") is the file-naming system of the Internet. URLs allow you to find any resource on the Internet. Examples of resources include text files, Web pages, program files, and audio and movie clips—in short, anything you might also find on your own computer. URLs allow you to find resources located on somebody else's computer. Table 4–1 shows several examples of typical URLs:

Table 4–1: *Typical URL Examples*

URL Example	Meaning
http://www.autodesk.com	Autodesk primary Web site
news://adesknews.autodesk.com	Autodesk news server
ftp://ftp.autodesk.com	Autodesk FTP server
http://www.autodeskpress.com	Autodesk Press Web site
http://www.upfrontezine.com	Editor Ralph Grabowski's Web site

Note that the http:// prefix is not required. Most of today's Web browsers automatically add the *routing* prefix, which saves you a few keystrokes. (The :// characters indicate a network address.)

LAUNCHING A WEB BROWSER (BROWSER COMMAND)

The Browser command starts a Web browser from within AutoCAD. Common Web browsers include AOL Netscape, Microsoft Internet Explorer, and Operasoft Opera.

The Browser command uses your computer's registered Web browser program. AutoCAD prompts you for the URL, such as http://www.autodesk.com. The Browser command can be used in scripts, toolbar or menu macros, and AutoLISP routines to access the Internet automatically, as shown in the following example:

```
Command: browser
Enter Web location (URL)  <http://www.autodesk.com>: (Enter the URL.)
```

The default URL is Autodesk's home page. After you type the URL and press Enter, AutoCAD launches the Web browser and contacts the Web site (see Figure 4–1).

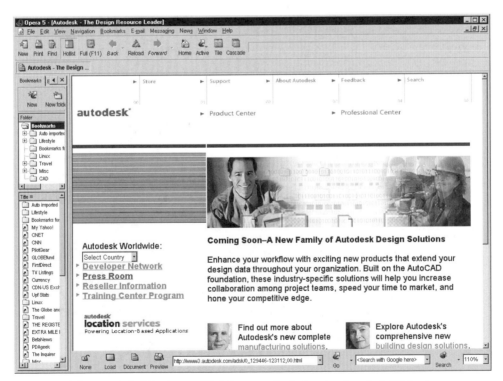

Figure 4–1 *The Opera Web browser displaying the Autodesk Web site.*

CHANGING THE DEFAULT WEB SITE

To change the default Web page that your browser starts with, change the INET-LOCATION system variable, which stores the URL used by the Browser command and the Browse the Web dialog box:

```
Command: inetlocation

Enter new value for INETLOCATION <"http://www.autodesk.com">: (Enter the URL.)
```

DRAWINGS ON THE INTERNET

When a drawing is stored on the Internet, you access it from within AutoCAD using the standard Open, Insert, and Save commands. Instead of specifying the file's location with the usual drive\folder\file name format, as in c:\autocad2002\filename.dwg, use the URL format. (Recall that the URL is the Internet's universal file-naming system used to access any file located on any computer hooked up to the Internet.)

OPENING DRAWINGS FROM THE INTERNET (OPEN COMMAND)

To open a drawing from the Internet (or your firm's Intranet), use the Open command (choose File>Open). Notice the Search the Web button to the right of the Look in drop-down list (see Figure 4–2).

When you choose the Search the Web button, AutoCAD opens the Browse the Web window, a simple Web browser that lets you to browse files at a Web site.

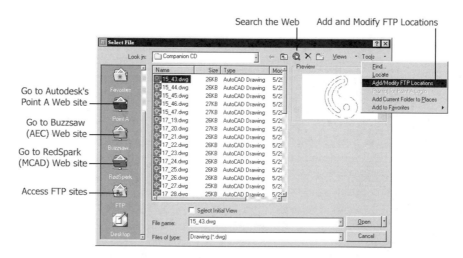

Figure 4–2 *The Select File dialog box's Internet features.*

By default, the Browse the Web dialog box displays the contents of the URL stored in the INETLOCATION system variable. You can change this to another folder or Web site, as noted earlier. Along the top, the dialog box has six buttons:

- **Back**—Returns to the previous URL.

- **Forward**—Goes forward to the next URL.

- **Stop**—Halts displaying the Web page (useful if the connection is slow or the page is very large).

- **Refresh**—Redisplays the current Web page.

- **Home**—Returns to the location specified by INETLOCATION.

- **Favorites**—Lists stored URLs (hyperlinks) or bookmarks. If you have previously used Internet Explorer, you will find all your favorites listed here. Favorites are stored in the \Windows\Favorites folder on your computer.

The Look in field allows you to type the URL. Alternatively, click on the down arrow to select a previous destination. If you have stored Web site addresses in the Favorites folder, then select a URL from that list.

You can either double-click on a file name in the window or enter a URL in the File Name field. The Table 4–2 provides examples of URLs you can type to open a drawing file.

When you open a drawing from the Internet, it will probably take much longer than opening a file on your computer. During the file transfer, AutoCAD displays a dialog box to report the progress. If your computer uses a 28.8 Kbps modem, you should allow about five to ten minutes per megabyte of drawing file. If your computer has access to a faster, DSL or cable connection to the Internet, you should expect a transfer speed of about a half-minute per megabyte.

Table 4–2: *Examples of URLs With Drawing Files*

Drawing Location	URL Example
Web or HTTP site	http://servername/pathname/filename.dwg
FTP site	ftp://servername/pathname/filename.dwg
Local file	drive:\pathname\filename.dwg
Network file	\\localhost\drive:\pathname\filename.dwg

It may be helpful to understand that the Open command does not copy the file from the Internet location directly into AutoCAD. Instead, it copies the file from the Internet to your computer's designated temporary subdirectory, such as C:\Windows\Temp (and then loads the drawing from the hard drive into AutoCAD). This is known as *caching*. It helps to speed up the processing of the drawing, because the drawing file is now located on your computer's fast hard drive, instead of on the relatively slow Internet.

TUTORIAL

The Autodesk Press Web site has an area that allows you to practice using the Internet with AutoCAD. In this tutorial you will open a drawing file located on that Web site.

1. Start AutoCAD.

2. Ensure that you have a live connection to the Internet. If you normally access the Internet via a telephone (modem) connection, dial your Internet service provider now.

3. From the menu bar, choose File>Open, or choose the Open icon on the toolbar. Notice that AutoCAD displays the Select File dialog box.

4. Choose the Search the Web button. Notice that AutoCAD displays the Browse the Web dialog box (Figure 4–3).

5. In the Look in field, type **practicewrench.autodeskpress.com** and press Enter. After a few seconds the Browse the Web dialog box displays the Web site.

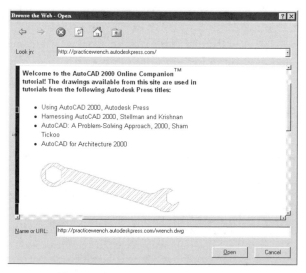

Figure 4–3 *Browse the Web window.*

6. In the Name or URL field, type **wrench.dwg** and press Enter. AutoCAD begins transferring the file (Figure 4–4). Depending on the speed of your Internet connection, this will take between a couple of seconds and a half-minute.

Notice the drawing of the wrench in AutoCAD, as shown in Figure 4–5.

AUTODESK WEB SITES

The Select File dialog box provides access to some of Autodesk's Web sites:

- **Point A**—A "portal" site that provides information about CAD as well as updates and bug patches for Autodesk software

- **Buzzsaw**—A project management and remote printing services site for the A/E/C industry (architecture, engineering, construction) (Figure 4–6)

- **RedSpark**—An e-commerce site for the manufacturing industry

Figure 4–4 *File Download dialog box.*

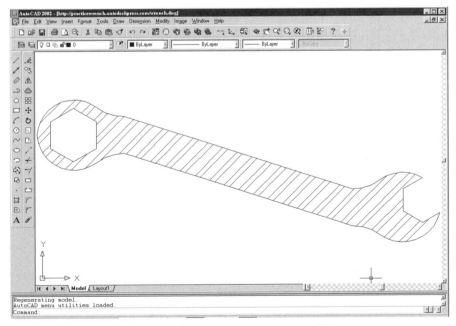

Figure 4–5 *Wrench drawing in AutoCAD.*

Figure 4–6 *Buzzsaw Web site.*

INSERTING A BLOCK FROM THE INTERNET (INSERT COMMAND)

When a block (symbol) is stored on the Internet, you can access it from within AutoCAD using the Insert command. When the Insert dialog box appears, choose the Browse button to display the Select Drawing File dialog box.

After you click the file name to select the file, AutoCAD downloads the file and continues with the Insert command's familiar prompts.

The process is identical for accessing external reference (xref) and raster image files. Other files that AutoCAD can access over the Internet include 3D Studio, SAT (ACIS solid modeling), DXB (drawing exchange binary), and WMF (Windows metafile). All of these options are found on the Insert menu on the menu bar.

ACCESSING OTHER FILES ON THE INTERNET

Most other file-related dialog boxes allow you to access files from the Internet or Intranet. This allows your firm or agency to have a central location that stores drawing standards. When you need to use a linetype or hatch pattern, for example, you access the LIN or PAT file over the Internet. More than likely you would have the location of those files stored in the Favorites list.

Some examples of file types that you can access over the Internet include the following:

- **Linetypes**—From the menu bar, choose Format>Linetype. In the Linetype Manager dialog box, choose the Load File and Look in Favorites buttons.

- **Hatch Patterns**—Use the Web browser to copy PAT files from a remote location to your computer.

- **Multiline Styles**—From the menu bar, choose Format>Multiline Style. In the Multiline Styles dialog box, choose the Load File and Look in Favorites buttons.

- **Layer Name**—From the menu bar, choose Express>Layers>Layer Manager. In the Layer Manager dialog box, choose the Import and Look in Favorites buttons.

- **LISP and ARX Applications**—From the menu bar, choose Tools>Load Applications.

- **Scripts**—From the menu bar, choose Tools>Run Scripts.

- **Menus**—From the menu bar, choose Tools>Customize Menus. In the Menu Customization dialog box, choose the Browse and Look in Favorites buttons.

- **Images**—From the menu bar, choose Tools>Displays Image>View.

You cannot access text files, text fonts (SHX and TTF), color settings, lineweights, dimension styles, plot styles, OLE objects, or named UCSs over the Internet.

SAVING THE DRAWING TO THE INTERNET (SAVE COMMAND)

When you are finished editing a drawing in AutoCAD, you can save it to a file server on the Internet with the Save command. If you inserted the drawing from the Internet (using the Insert command) into the default drawing1.dwg drawing, AutoCAD insists that you first save the drawing to your computer's hard drive.

When a drawing of the same name already exists at that URL, AutoCAD warns you, just as it does when you use the SaveAs command. Recall from the discussion of the Open command that AutoCAD uses your computer system's Temporary subdirectory, hence the reference to it in the dialog box.

USING HYPERLINKS WITH AUTOCAD (HYPERLINK COMMAND)

AutoCAD allows you to employ hyperlinks (URLs) in two ways: directly within the AutoCAD drawing, and indirectly in DWF files displayed by a Web browser.

HYPERLINKS INSIDE AUTOCAD

AutoCAD allows you to add hyperlinks to any object in the drawing. You can attach one or more hyperlinks to one or more objects.

Determine whether an object has a hyperlink by passing the cursor over it. If it does, the cursor will display the "linked Earth" icon as well as a ToolTip describing the link, as shown in Figure 4–7.

Figure 4–7 *The cursor reveals a hyperlink.*

If for some reason you do not want to see the hyperlink cursor, you can turn it off. From the menu, choose Tools>Options, and then click on the User Preferences tab. The Display Hyperlink Cursor and Shortcut Menu item toggles the display of the hyperlink cursor as well as the Hyperlink ToolTip option on the cursor menu.

Hyperlinks are created, edited, and removed with the Hyperlink command (which displays a dialog box) and the -Hyperlinks command (for prompts at the command line). The Hyperlink command (Insert>Hyperlink) prompts you to "Select objects" and then displays the Insert Hyperlink dialog box, as shown in Figure 4–8. (As a shortcut, you can press Ctrl+K or choose the Insert Hyperlink button on the toolbar.)

As an alternative to the menu command, use the -Hyperlink command. This command displays its prompts at the command line and is useful for scripts and AutoLISP routines. The -Hyperlink command has the following syntax:

```
Command: -hyperlink
Enter an option [Remove/Insert] <Insert>: (Press Enter.)
Enter hyperlink insert option [Area/Object] <Object>: (Press Enter.)
Select objects: (Pick an object.)
1 found Select objects: (Press Enter.)
Enter hyperlink <current drawing>: (Enter the name of the document or Web site.)
Enter named location <none>: (Enter the name of a bookmark or AutoCAD view.)
Enter description <none>: (Enter a description of the hyperlink.)
```

The command also allows you to remove a hyperlink. It does not, however, allow you to edit a hyperlink. To edit the hyperlink, use the Insert option to respecify the hyperlink.

In addition, the -Hyperlink command allows you to create a hyperlink *area*—a rectangular area that can be thought of as a 2D hyperlink, as shown in Figure 4–9. (The dialog box-based Hyperlink command does not create hyperlink areas.) When you choose the Area option, the rectangle is placed automatically on layer URLLAYER and colored red.

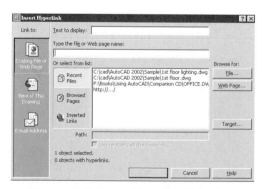

Figure 4–8 *Insert Hyperlink dialog box.*

Figure 4–9 *A rectangular hyperlink area.*

In the following sections you will learn how to apply and use hyperlinks in an AutoCAD drawing and in a Web browser via the dialog box–based Hyperlink command.

TUTORIAL

Let us see how hyperlinks function by working through an example. In this case we have the drawing of a floor plan. We will add hyperlinks to another AutoCAD drawing, a Microsoft Word document, and a Web site.

Hyperlinks must be attached to objects. For this reason, we will place text in the drawing and then attach the hyperlinks to the text.

1. Start AutoCAD.

2. Open the 1st floor plan.dwg file found in the AutoCAD 2002\Sample folder. (If necessary, click on the Model tab to display the drawing in model space.)

3. Start the Text command and enter text into the drawing as follows (see Figure 4–10):

    ```
    Command: text

    Current text style: "Standard" Text height: 0.20

    Specify start point of text or [Justify/Style]: (Select a
    point in the drawing.)

    Specify height <0.2000>: 2

    Specify rotation angle of text <0>: (Press Enter.)

    Enter text: Site Plan

    Enter text: Lighting Specs

    Enter text: Electrical Bylaw

    Enter text: (Press Enter.)
    ```

4. Click the "Site Plan" text to select it.

5. From the menu, select Insert>Hyperlink. Notice the Insert Hyperlink dialog box (Figure 4–11).

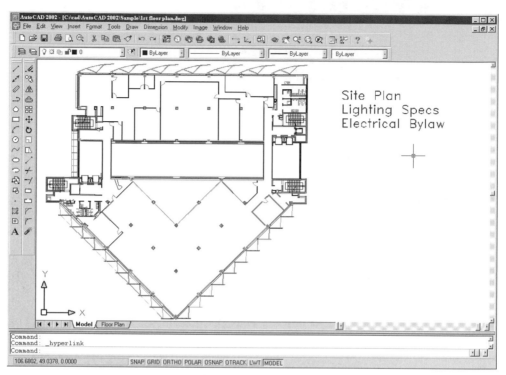

Figure 4–10 *Text placed in the drawing.*

Figure 4–11 *Insert Hyperlink dialog box.*

6. Below Browse for, choose the File button. Notice the Browse the Web - Select Hyperlink dialog box.

7. Go to AutoCAD 2002's Sample folder and select the City base map.dwg file. Choose Open.

Notice that AutoCAD does not open the drawing; rather, it copies the file's name to the Insert Hyperlinks dialog box. You can fill in two other fields:

- **Text to Display**—Provides a description for the hyperlink, which is displayed by the ToolTip. If you leave this blank, the URL is displayed by the ToolTip. For this tutorial, enter the description as follows:

  `Text to display:` **`Click to view the site plan`**

- **Target**—Provides the location for the hyperlinked file (when opened) and is sometimes called a "Bookmark." In AutoCAD, the bookmark is a named view (created with the View command) or a layout. Click on Target and select "ANSI D Plot" (see Figure 4–12).

8. Choose OK to dismiss the dialog box.

    Move the cursor over the Site Plan text. Notice the display of the "linked Earth" icon; a moment later, the ToolTip displays the text "City base map.dwg" (Figure 4–13).

9. Repeat the Hyperlink command twice more, attaching the files shown in Table 4–3 to the listed drawing text.

    You have now attached a drawing, a text document, and a Web document to objects in the drawing. Now you will try out the hyperlinks.

Table 4–3: *Files to Hyperlink*

Text	URL
Lighting Specs	Convert.Txt (found in AutoCAD 2002\sample\vba)
Electrical Bylaw	Cadmgr.htm (also found in AutoCAD 2002)

Figure 4–12 *Selecting a target file.*

10. Click "Site Plan" to select it. Right-click and choose Hyperlink>Open "City base map.dwg" from the cursor menu (Figure 4–14). Notice that AutoCAD opens City base map.dwg.

   To see both drawings, choose Window>Tile Vertically from the menu bar (Figure 4-15).

11. Select and then right-click on the "Lighting Specs" hyperlink. Choose Hyperlink>Open. Notice that Windows starts a word processor and opens the Convert.Txt file.

Figure 4–13 *Hyperlink cursor and ToolTip.*

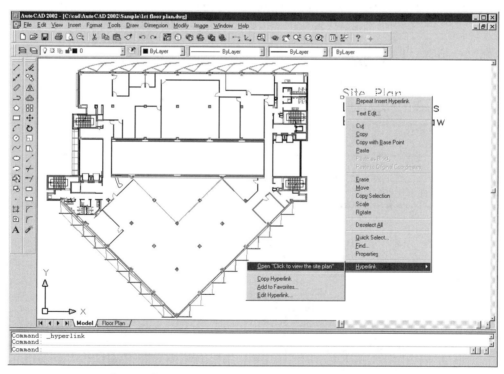

Figure 4–14 *Selecting a hyperlink from the cursor menu.*

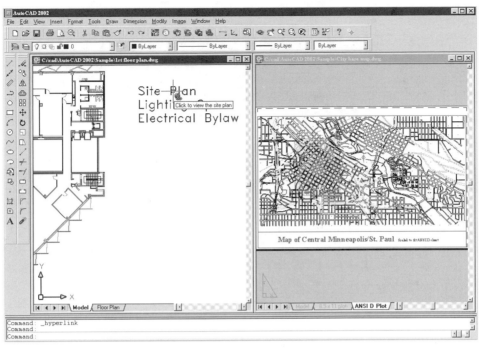

Figure 4–15 *Viewing two drawings.*

12. If your word processor has a Web toolbar, open it. Choose the Back button (the back arrow), as shown in Figure 4–16. Notice how doing so sends you into AutoCAD.

13. Select and then right-click on the "Electrical Bylaw" hyperlink. Choose Hyperlink>Open. Notice that Windows starts your Web browser and opens the Cadmgr.htm file (Figure 4–17).

14. Back in AutoCAD, right-click on any toolbar and choose Web. Notice that the Web toolbar has four buttons, as shown in Figure 4–18. (See Table 4–4.)

Table 4–4: *Web Toolbar Buttons*

Button	Command	Meaning
Go Back	HyperlinkBack	Moves back to the previous hyperlink
Go Forward	HyperlinkFwd	Moves to the next hyperlink
Stop Navigation	HyperlinkStop	Stops the hyperlink access action
Browse the Web	Browser	Launches the Web browser

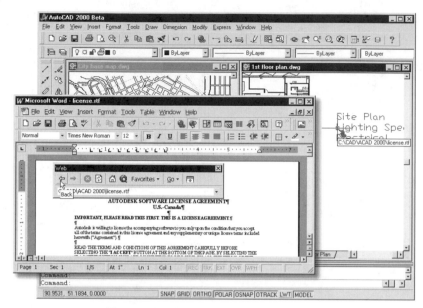

Figure 4–16 *Viewing a Word document.*

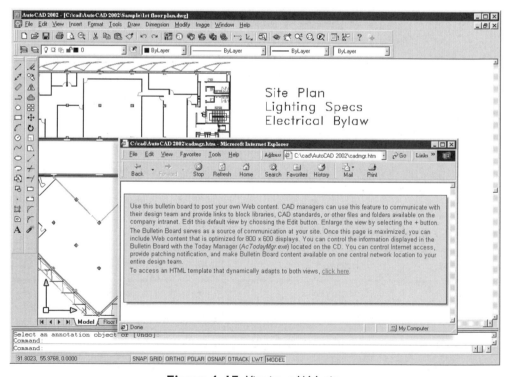

Figure 4–17 *Viewing a Web site.*

Figure 4–18 *AutoCAD's Web toolbar.*

15. Try choosing the Go Forward and Go Back buttons. Notice how these allow you to navigate among the drawings, the RTF document, and the Web page.

When you work with hyperlinks in AutoCAD, you might come across these limitations:

- AutoCAD does not check that the URL you enter is valid.

- If you attach a hyperlink to a block, the hyperlink data is lost when you scale the block unevenly, stretch the block, or explode it.

- Wide polylines and rectangular hyperlink areas are only "sensitive" on their outline.

Pasting Text as a Hyperlink

AutoCAD 2002 has a shortcut for creating hyperlinks in the drawing. The undocumented PasteAsHyperlink command pastes any text from the Clipboard as a hyperlink to any object in the drawing. The steps are as follows:

1. In a word processor, select some text and copy it to the Clipboard (via Ctrl+c or the Edit>Copy command). The text can be a URL (such as http:// www.autodeskpress.com) or any other text.

2. Switch to AutoCAD and choose Edit>Paste As Hyperlink from the menu bar. Note that this command does not work (is grayed out) if anything else, such as a picture, is in the Clipboard.

3. Select one or more objects as prompted:

    ```
    Command: _pasteashyperlink
    Select objects: (Pick an object.)
    1 found Select objects: (Press Enter.)
    ```

4. Pass the cursor over the object and note the hyperlink cursor and ToolTip. The ToolTip displays the same text that you copied from the document.

If the text you copy to the Clipboard is very long, AutoCAD displays only portions of it in the ToolTip, using ellipses (...) to shorten the text.

You cannot select ordinary text in the AutoCAD drawing to paste as a hyperlink. The MatchProp command does not work with hyperlinks.

You can, however, copy the hyperlink from one object to another. To do so, select the object, right-click, and then choose Hyperlink>Copy Hyperlink from the cursor menu. The hyperlink is copied to the Clipboard. You can now paste the hyperlink into another document or use AutoCAD's Edit>Paste as Hyperlink command to attach the hyperlink to another object in the drawing.

Highlighting Objects with URLs

Although you can see the rectangle of area URLs, the hyperlinks themselves are invisible. For this reason, AutoCAD has the QSelect command, which highlights all objects that match specifications. The steps are as follows:

1. From the menu bar, choose Tools>Quick Select. AutoCAD displays the Quick Select dialog box (Figure 4-19).

2. Enter specifications into the fields as follows:

   Apply to: **Entire drawing**

   Object Type: **Multiple**

   Properties: **Hyperlink**

   Operator: * **Wildcard Match**

   Value: *

3. Choose OK. AutoCAD highlights all objects that have a hyperlink. Depending on your computer's display system, the highlighting shows up as dashed lines or as another color.

Figure 4–19 *Quick Select dialog box.*

Editing Hyperlinks

Now that you know where the objects with hyperlinks are located, you can use the Hyperlink command to edit their hyperlinks and related data. Select the hyperlinked object and start the Hyperlink command (press Ctrl+K). When the Edit Hyperlink dialog box appears (it looks identical to the Insert Hyperlink dialog box), make the changes and choose OK.

Removing Hyperlinks from Objects

To remove a URL from an object, use the Hyperlink command. When the Edit Hyperlink dialog box appears, choose the Remove Hyperlink button.

To remove a rectangular area hyperlink, use the Erase command. When you select the rectangle, AutoCAD erases the rectangle. As an alternative, use the -Hyperlink command's Remove option as follows:

```
Command: -hyperlink

Enter an option [Remove/Insert] <Insert>: r

Select objects: all

1 found Select objects: (Press Enter.)

1. www.autodesk.com

2. www.autodeskpress.com

Enter number, hyperlink, or * for all: 1

Remove, deleting the Area.

1 hyperlink deleted.
```

HYPERLINKS OUTSIDE AUTOCAD

Hyperlinks you place in the drawing are also available for use outside of AutoCAD. The Web browser uses hyperlink(s) when the drawing is exported in DWF format. To help make the process clearer, follow these steps:

1. Open a drawing in AutoCAD.

2. Attach hyperlinks to objects in the drawing with the -Hyperlinks command. (To attach hyperlinks to areas, use the -Hyperlink command's Area option.)

3. Export the drawing in DWF format using the Plot command's "DWF ePlot PC2" plotter configuration.

4. Copy the DWF file to your Web site.

5. Start your Web browser with the Browser command.

6. View the DWF file and click on a hyperlink spot.

THE DRAWING WEB FORMAT

To display AutoCAD drawings on the Internet, Autodesk invented the "drawing Web format" (DWF). The DWF file has several benefits and some drawbacks, compared to DWG files.

The DWF file is compressed to make it as much as eight times as small as the original DWG drawing file, so that it takes less time to transmit over the Internet, which is particularly vital with relatively slow telephone modem connections. The DWF format is more secure because the original drawing is not being displayed; another user cannot tamper with the original DWG file.

However, the DWF format has some drawbacks:

- You must go through the extra step of translating from DWG to DWF.
- DWF files cannot display rendered or shaded drawings.
- DWF is a flat, 2D file format; therefore, it does not preserve 3D data, although you can export a 3D view.
- AutoCAD itself cannot display DWF files.
- DWF files cannot be converted back to DWG format without the use of file translation software from a third-party vendor.
- Earlier versions of DWF did not handle paper space objects (version 2.x and earlier), or linewidths and non-rectangular viewports (version 3.x and earlier).

To view a DWF file on the Internet, your Web browser needs a *plug-in*—a software extension. The DWF plug-in should have been added to your browser when AutoCAD was installed on your computer. Autodesk also makes the DWF plug-in freely available from its Web site at http://www.autodesk.com/whip. It is a good idea to check regularly for updates to the DWF plug-in, which is updated about twice a year.

Autodesk also provides two other options for viewing DWF files. CADViewer Light is designed to work on all operating systems and computer hardware because it is written in Java. As long as your Windows, Macintosh, or Linux computer has access to Java (which is included with most Web browsers), it can view DWF files. Volo View Express is a standalone viewer that views and prints DWG, DWF, and DXF files. Both products can be downloaded free from the Autodesk Web site.

CREATING A DWF FILE

To create a DWF file from AutoCAD, follow these steps:

1. Enter the Plot command or choose File>Plot from the menu bar. Notice that AutoCAD displays the Plot dialog box.

2. In the Name List box (found in the Plotter Configuration area), choose "PublishToWeb DWF pc3" (Figure 4–20).

3. Choose the Properties button. Notice the Plotter Configuration Editor dialog box.

4. Choose Custom Properties in the tree view. Notice the Custom Properties button, which appears in the lower half of the dialog box (see Figure 4–21).

5. Choose the Custom Properties button. Notice the DWF Properties dialog box (Figure 4–22).

Resolution Area

Unlike AutoCAD DWG files, which are based on real numbers, DWF files are based on integers. The Medium precision setting saves the drawing using 16-bit integers, which is adequate for all but the most complex drawings. High resolution saves the DWF file using 20-bit integers, while Extreme resolution saves the file using 32-bit integers. Figure 4–23 shows an extreme closeup of a drawing exported in DWF format. The portion on the left was saved at medium resolution and shows some bumpiness in the arcs. The portion on the right was saved at extreme resolution.

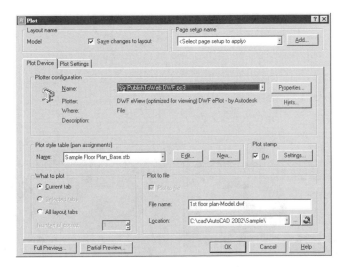

Figure 4–20 *Plot dialog box.*

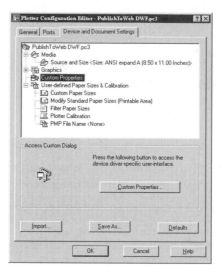

Figure 4–21 *Plotter Configuration Editor dialog box.*

Figure 4–22 *DWF Properties dialog box.*

Figure 4–23 *DWF output in medium resolution (left) and extreme resolution (right).*

More significant is the difference in file size. When I created a DWF file from the 1st floor plan.dwg file (342 KB), the medium-resolution DWF file was just 27 KB, while the extreme-resolution file became 1,100 KB—about forty times as large. That means that the medium-resolution DWF file gets transmitted over the Internet forty times as fast, a significant savings in time.

Format Area

Compression further reduces the size of the DWF file. You should always use compression, unless you know that another application cannot decompress the DWF file. Compressed binary format is seven times as small as ASCII format (7,700 KB). Again, that means the compressed DWF file gets transmitted over the Internet seven times as fast as the ASCII DWF file.

Other Options

The following additional options are also available. In most cases, you would turn on all options.

- **Background Color Shown in Viewer**—Allows you to choose any of AutoCAD's 255 color, although white is probably the best background color

- **Include Layer Information**—Includes layers, allowing you to toggle layers off and on when the drawing is viewed in the Web browser

- **Include Scale and Measurement Information**—Allows you to use the Location option in the Web browser WHIP! plug-in to show scaled coordinate data

- **Show Paper Boundaries**—Includes a rectangular boundary at the drawing's extents

- **Convert .DWG Hyperlink Extensions to .DWF**—Includes hyperlinks in the DWF file

6. Choose OK to exit the dialog boxes, and then return to the Plot dialog box.

7. Accept the DWF file name listed in the File Name text box, or type a new name. If necessary, change the location at which the file will be stored. Note the two buttons: one has an ellipsis (**...**) and displays the Browse for Folder dialog box. The second button brings up AutoCAD's internal Web browser.

8. Choose OK to save the drawing in DWF format.

VIEWING DWF FILES

To view a DWF file you must use a Web browser with a special plug-in that allows the browser to interpret the file correctly. (Remember: You cannot view a DWF file in AutoCAD.) Autodesk has named their DWF plug-in WHIP!, which is short for "Windows HIgh Performance."

Autodesk updates the DWF plug-in approximately twice a year. Each update includes some new features. The functions of the DWF plug-in are summarized as follows:

- View (in a browser) DWF files created by AutoCAD.

- Right-click on the DWF image to display a cursor menu with commands.

- Use the real-time pan and zoom features to change the view of the DWF file just as you can change a view in an AutoCAD drawing file.

- Use embedded hyperlinks to display other documents and files.

- Compress files—the DWF file appears in your Web browser much faster than the equivalent DWG drawing file would.

- Print the DWF file alone or along with the entire Web page.

- View drawings using either Netscape Communicator/Navigator or Microsoft Internet Explorer. A separate browser-specific plug-in is required, depending on which browser you use.

- Drag and drop a DWG file from a Web site into AutoCAD as a new drawing or as a block.

- View a named view stored in the DWF file.

- Specify a view using x, y coordinates.

- Toggle layers off and on.

If the plug-in is not installed or is an older version, then you need to download it from Autodesk's Web site at http://www.autodesk.com/whip (Figure 4–24).

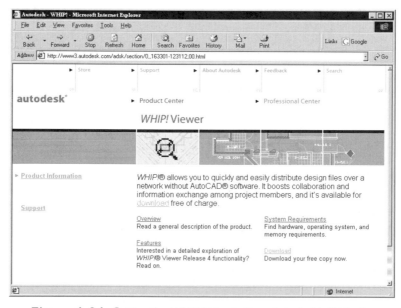

Figure 4–24 *Getting the WHIP! plug-in from Autodesk's Web site.*

DWF PLUG-IN COMMANDS

To let others view your DWF file over the Internet, you need to embed the DWF file in a Web page. Use the PublishToWeb command, as described in Chapter 3.

To display the DWF plug-in's commands, position the cursor over the DWF image in the Web browser and click the right mouse button. This displays a cursor menu with commands such as Pan, Zoom, and Named Views (Figure 4-25). To choose a command, place the cursor over the command name and click the left mouse button. The commands are described as follows:

- **Pan**—Pans the view around in real time. Click the left mouse button and move the mouse. The cursor changes to an open hand, signaling that you can pan the view around the drawing. Naturally, panning only works when you are zoomed in; it does not work in full-view mode.

- **Zoom**—Is like the Zoom command in AutoCAD. The cursor changes to a magnifying glass. Hold down the left mouse button and move the cursor up (to zoom in) and down (to zoom out).

- **Layers**—Displays a non-modal dialog box that lists all layers in the drawing. (A *non-modal* dialog box remains on the screen; unlike with AutoCAD's modal dialog boxes, you do not need to dismiss a non-modal dialog box to continue

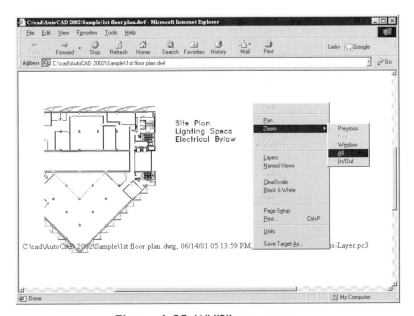

Figure 4–25 *WHIP!'s cursor menu.*

working.) Click a layer name to toggle its visibility between on (yellow light bulb icon) and off (blue light bulb), as shown in Figure 4–26.

- **Named Views**—Displays a non-modal dialog box that allows you to select a named view. Click a named view to see it, as shown in Figure 4–27. Click the small x in the upper-right corner to dismiss the dialog box.

- **Clear Scale**—Changes the drawing's colors to grayscale and makes the background white.

- **Black & White**—Changes the drawing's colors to black and the background to white.

- **Units**—Allows you to select units for the drawing (inches, meters, points, and so on) and separate units for fonts and lineweights.

- **Print**—Prints the DWF image alone. To print the entire Web page (including the DWF image), use the browser's Print button.

- **Save Target As**—Saves the DWF file to your computer's hard drive. You cannot use the Save As command until the entire DWF file has been transmitted to your computer.

Figure 4–26 *WHIP!'s Layers dialog box.*

Figure 4–27 *WHIP!'s Named Views dialog box.*

SUMMARY

In this chapter you learned how to use AutoCAD's Internet-based features. You launched a Web browser from AutoCAD, and learned about URLs (uniform resource locators). You opened and saved drawings to and from the Internet, and placed hyperlinks in a drawing. You also created DWF files from drawings, and viewed the DWFs using a Web browser. Through this chapter, you learned how AutoCAD helps you integrate your drawings into a web-based environment.

SECTION

III

Customizing
AutoCAD

Customizing without Programming

One of the primary reasons AutoCAD has retained its popularity over the years is its ability to be easily personalized to fit the needs of a large variety of users. Even users with little or no sophisticated programming skills can customize an out-of-the-box copy of AutoCAD to an amazing degree.

One of the best and easiest ways to increase significantly your AutoCAD productivity is through customization. It does not take long to acquire the knowledge and master the skills required to perform basic customization of menus, for example, and in a relatively short time you can tailor AutoCAD to meet your specific work requirements.

This chapter will focus upon the simplest—and perhaps most productive—of customizing techniques: streamlining the accessing of AutoCAD commands.

This chapter will cover the following topics:

- Creating command aliases
- Customization and menu files
- Customizing toolbars
- Adding an item to the cursor menu

CREATING COMMAND ALIASES

The most basic and simplest means of customizing AutoCAD is through the creation of *command aliases*. A command alias is a one-, two-, or three-key combination that launches a standard AutoCAD command when you type the combination on the keyboard and then press Enter. They are great timesavers. Pressing the "RE" key combination and then Enter to initiate the REGEN command, for example, is obviously much faster than typing **regen** and then pressing Enter. For most users it is even faster than using the REGEN command item on the View menu—assuming that you know the REGEN command resides on the View menu.

In fact, AutoCAD 2000/2000i/2002 come "out-of-the-box" with well over two hundred of these command aliases predefined and ready to be used. This number includes many of AutoCAD's system variables, which are also assigned aliases. Veteran AutoCAD users generally opt for typing **H** for the BHATCH command, **C** for the CIRCLE command, **E** for the ERASE command, etc. Assuming that one hand is always free for keyboard input, these aliases can significantly increase your productivity.

Command aliases are defined and stored in a file with the name ACAD.PGP. In a standard AutoCAD installation, this file is installed under the AutoCAD\ folder. The "PGP" extension is an abbreviation for ProGram Parameters, and the file is frequently referred to as "the PGP file." This file is a standard ASCII text file that can be easily edited in any Windows text editor, such as Notepad.

The ACAD.PGP file is actually divided into two main sections: the External Command section and the Command Alias section. In a modern Windows-type environment, the External Command section is largely anachronistic. This chapter will look at the Command Alias section. A portion of the Command Alias section from AutoCAD 2002's PGP file is shown here.

Note: Some entries have been removed from the following list.

```
; [AutoCAD Command Aliases]
3A,         *3DARRAY
3DO,        *3DORBIT
3P,         *3DPOLY
A,          *ARC
AEX,        *DBCONNECT
AL,         *ALIGN
AP,         *APPLOAD
AR,         *ARRAY
-AR,        *-ARRAY
```

```
ATE,       *ATTEDIT
-ATE,      *-ATTEDIT
ATT,       *ATTDEF
ATTE,      *-ATTEDIT
B,         *BLOCK
-B,        *-BLOCK
BH,        *BHATCH
```

A short inspection of this excerpt shows the basic scheme of defining a command alias:

```
<Alias>,   *<Full command name>
```

The <Alias> represents the one-, two-, or three-letter keyboard combination that can be typed as a substitute for the official command name. A comma (,) must immediately follow the alias. Next you enter the full name of the AutoCAD command for which the alias will stand. This name must be immediately preceded by an asterisk (*). There can be any convenient number of spaces between the comma and the asterisk. These spaces are inserted to separate visually the alias from its command so that the list is easier to read.

Although the predefined aliases are quite extensive, you may want to edit a predefined alias to suite your own needs or add an alias for a command or system variable not included in the predefined list. In the following exercise you will add an alias for the CURSORSIZE system variable and edit the alias for the 3DARRAY command.

Note: As with any of AutoCAD's customization files, you should make a backup copy of the ACAD.PGP file before loading it into a text editor. You can always revert back to the original if anything goes wrong.

EDITING AND ADDING A COMMAND ALIAS TO THE PGP FILE

1. Start Windows Notepad and navigate to the Acad2002\ folder. Find and load the ACAD.PGP file. The top of the file should resemble Figure 5–1.

2. Scroll down to the [AutoCAD Command Aliases] section and then find the entry for the 3DARRAY command. In an unmodified PGP file it will be the first entry.

3. For the purpose of this exercise, assume you want to change the alias for the 3DARRAY command from its current 3A to 3DA. Use the text editor to make this change.

4. Scroll down the list of aliases to find the alias CP, for the COPY command. Use the text editor to insert the following alias:

```
CS,      *CURSORSIZE
```

```
; [Operating System Commands]

CATALOG,    DIR /W,          8,*File specification:,
DEL,        DEL,             8,*File to delete:,
DIR,        DIR,             8,*File specification:,
EDIT,       START EDIT,      9,*File to edit:,
EXPLORER,   START EXPLORER,  1,,
NOTEPAD,    START NOTEPAD,   1,*File to edit:,
PBRUSH,     START PBRUSH,    1,,
SH,         ,                1,*OS Command:,
SHELL,      ,                1,*OS Command:,
START,      START,           1,*Application to start:,
TYPE,       TYPE,            8,*File to list:,


; [AutoCAD Command Aliases]

3A,         *3DARRAY
3DO,        *3DORBIT
3F,         *3DFACE
3P,         *3DPOLY
A,          *ARC
AA,         *AREA
AAD,        *DBCONNECT
ADC,        *ADCENTER
AEX,        *DBCONNECT
```

Figure 5–1 *The ACAD.PGP file in a text editor.*

Note: Although it is not required, it is a good practice when adding lines to the PGP file to add them in alphabetical order, by alias. This way they are easier to find. Also, the PGP file is case insensitive, but it is good practice to use all UPPER CASE for uniformity.

5. This portion of the PGP file should now resemble Figure 5–2 (in which bold has been added for emphasis). Save the file and exit the text editor.

Note: AutoCAD automatically reads and loads the PGP file each time it is started. If you performed the above changes to the PGP file with AutoCAD running, you can perform the next step to force AutoCAD to reread and update the file with the changes. Otherwise, the altered PGP file will be read the next time AutoCAD is started.

6. To force AutoCAD to reread the PGP file, run the REINIT (for reinitialize) command by typing REINIT and pressing Enter. In the Re-initialization dialog box (shown in Figure 5–3), select PGP File and press OK.

7. Test your changes in AutoCAD. Type **3DA** and press Enter. AutoCAD should start the 3DARRAY command. Press Esc to abort the command. Now type **CS** and press Enter. AutoCAD should prompt for the new cursor size. Press Enter to accept the current size.

CUSTOMIZATION AND MENU FILES

Before we move on to the customization of AutoCAD toolbars and pull-down menu items, the role of the various AutoCAD menu files needs to be understood.

```
C,              *CIRCLE
CH,             *PROPERTIES
-CH,            *CHANGE
CHA,            *CHAMFER
CO,             *COPY
COL,            *COLOR
COLOUR,         *COLOR
CS,             *CURSORSIZE
CP,             *COPY
D,              *DIMSTYLE
DAL,            *DIMALIGNED
```

Figure 5–2 *Adding a new alias to the PGP file.*

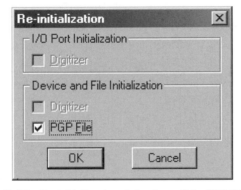

Figure 5–3 *The Re-initialization dialog box of the REINIT command.*

 Note: If you made an error in editing the PGP file—such as forgetting to place a comma after the alias or an asterisk before the command name—AutoCAD will report the error and the line number on which the error occurred on the command line. For example, the error message "Syntax error in acad.pgp file on line 90 in field 1" indicates that the comma following the alias was omitted.

 Note: Command aliases will only initiate a command; they cannot be used to specify command options. To automate both command initiation and subsequent options selection, you need to build simple AutoLISP keyboard macros. Such macros are discussed in Chapter 8 of this book, Introduction to AutoLISP Programming.

This is a topic of some confusion even among veteran AutoCAD users, and understandably so, because any menu loaded into AutoCAD can—and usually does—consist of not one but a "family" of closely related menu files. Some of these files automatically generate or update others of the family. The one item that this family of files has in common is the root name. The file extension varies, however.

Take the default out-of-the-box AutoCAD menu as an example. It consists of a group of five files, all with the root name *acad*. The full name and function of each of these files is as follows:

- **ACAD.MNU**—This is the so-called menu "template" file. It is a plain-text ASCII file, readable by any standard text editor application, such as Notepad. It contains the functional menu code along with non-functional comments.

- **ACAD.MNS**—This the so-called menu "source" file. It, too, is a plain-text ASCII file and is generated by the ACAD.MNU file. If no menu or toolbar customization has taken place using the TOOLBAR command, it is identical to the ACAD.MNU file with nonfunctional comment lines stripped away.

- **ACAD.MNC**—This is the so-called "compiled" menu file. It is the file that is actually loaded into AutoCAD on startup. It is in an unreadable, machine code, compiled format. The ACAD.MNS file is used as the "source" for this compilation.

- **ACAD.MNR**—This is the "resource" menu file. Its sole purpose is to contain the definitions of the various icons bitmaps used by the AutoCAD toolbars.

- **ACAD.MNL**—This is the "LISP" menu file. It contains plain-text AutoLISP code used by items in the menu. This file many not be present with files other than the ACAD menu files.

The problem arises when the AutoCAD interface (i.e., the TOOLBAR command) is used to modify current toolbars or create new toolbars. These modifications are written to the ACAD.MNS file, not to the ACAD.MNU file. The ACAD.MNU file is "unaware" of any toolbar modifications made *via* the TOOLBAR command method. The TOOLBAR command is called by the Toolbars item on the View menu and by the Customize item on the context (right-click) menu displayed by right-clicking on any individual tool on a toolbar.

 Note: If you explicitly load the ACAD.MNU file using the MENU command after modifying or creating toolbars, a new ACAD.MNS file will be created, and any toolbar changes to the ACAD.MNS file will be overwritten and lost. Fortunately, AutoCAD issues a warning to this effect under these circumstances.

There are two popular ways to circumvent this awkward situation:

- **Method 1** requires that you copy and paste any changes to the ***TOOLBARS section of the ACAD.MNS menu to the ACAD.MNU menu, in effect bringing the ACAD.MNU menu file "up to date."

- **Method 2** suggests that you disregard the ACAD.MNU altogether and either take it out of the AutoCAD search path or rename it and use the ACAD.MNS file exclusively for all menu customization work. Hiding or renaming the ACAD.MNU file will prevent it from being accidentally loaded, and overwriting toolbars changes in the current ACAD.MNS file. This method also keeps the

ACAD.MNU file available to regenerate a new ACAD.MNS file should the latter become accidentally lost or destroyed.

 Note: You can read more about the various AutoCAD menu files and the role they play in customization procedures in Chapter 7, Advanced Customization.

CUSTOMIZING AUTOCAD TOOLBARS

Modifying AutoCAD's toolbars or creating your own toolbars is another easy way to customize AutoCAD's working environment and increase your efficiency. AutoCAD's out-of-the-box floating toolbars are not unlike the toolbars found in many other modern Windows applications. They are great timesavers because they group a set of related AutoCAD commands together into a single unit that can be moved around the work area, reshaped, and even docked to the sides of the drawing editor. And, of course, they can be hidden away as well. Unlike the commands contained on AutoCAD's menus, the icons representing the toolbar's commands are always visible and easy to access.

There are several ways you can modify the 26 toolbars that are shipped with AutoCAD. You can rearrange the order in which the tools appear, you can eliminate tools that you seldom or never use, and you can add new tools. If you add new tools, they can be from an extensive standard "library" of tools and their associated icons, or you can make your own tools and design your own icons. Lastly, you can make completely new toolbars. Figure 5–4 shows a portion of AutoCAD 2002's standard screen layout with a floating toolbar displayed. The Draw, Modify, Object Properties, and Standard toolbars are shown in their default docked positions. The Solids toolbar is shown in a floating position in the drawing area.

MODIFYING AN EXISTING TOOLBAR

The ability to modify existing toolbars is a powerful customization feature. You may, for example, want to add a tool that you use frequently to an existing toolbar or remove a tool that you seldom use. Both operations are easy to perform and result in a customized toolbar that better suits your particular needs. In the following exercise you will add the PEDIT command to the Modify toolbar.

EXERCISE: ADDING A TOOL TO A TOOLBAR

1. With AutoCAD running, right-click on any toolbar icon. AutoCAD displays the toolbar status menu as shown in Figure 5–5.

2. Select Customize to display the Customize dialog box. This places AutoCAD in "customize" mode. If necessary, click on the Commands tab to make the Commands folder active (see Figure 5–6). The Categories window displays a list of the standard AutoCAD pull-down menu categories. The PEDIT command is

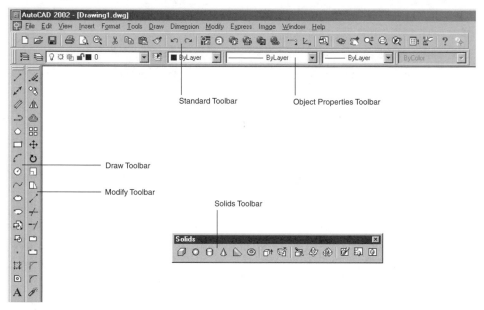

Figure 5–4 *AutoCAD 2002 with four docked and one floating toolbar displayed.*

Note: Modifying toolbars causes modifications to several menu files. It is advisable that you make copies of these files and place them in a separate folder just in case you need to revert to them. You should make copies of the following menu files: **.MNU, **.MNS, and **.MNR, where ** represents the root name of the menu with which you are working.

found on the Modify menu, so select Modify from the list. The library of Modify commands is displayed in the Commands window as shown in Figure 5–7.

3. Find and click on the Edit Polyline entry (at the bottom of the window). Note that for confirmation a description of the command is displayed under the Categories window.

4. Left-click and drag the PEDIT command's icon to the Modify toolbar. Note as you drag the icon that it changes to a generic shape and a small "+" sign appears in the corner.

5. As the icon touches the Modify toolbar, a thick, horizontal I-beam bar appears. This indicates where the icon will appear when you release the left mouse button.

6. While still pressing the left mouse button, position the I-beam bar between the FILLET and the EXPLODE commands' icons and release the left mouse button. The Modify toolbar should now resemble Figure 5–8.

7. As long as the Customize dialog box is open, AutoCAD is in the customize mode, and you can left-click and drag any toolbar icon to reposition it within its

Figure 5–5 *Right-clicking on any toolbar icon displays the toolbar status menu.*

toolbar. To delete an icon from a toolbar, drag and drop it into the drawing area. AutoCAD will issue the warning shown in Figure 5–9.

Using the preceding steps you can add or subtract any number of tools from any existing toolbar.

CREATING A NEW TOOLBAR

Creating new toolbars is another excellent way to increase your AutoCAD efficiency. You can create new toolbars containing any commands you wish. If you work with 3D solids, for example, you may want to create a new toolbar containing a combination of the most frequently used commands from both the Solids and the Solids Editing toolbars. Or you may want to make a new toolbar containing only the five or six dimension commands that you use.

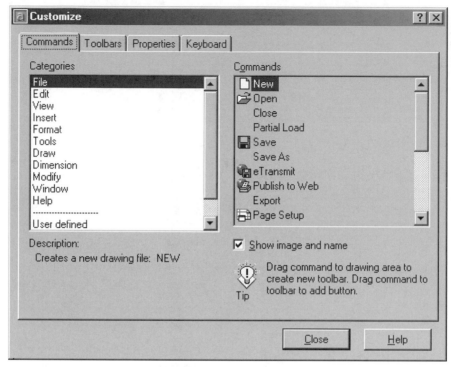

Figure 5–6 *The Customize dialog box places AutoCAD in customize mode.*

There are two ways to create a new toolbar using the Customize dialog box that you used in the preceding exercise. Both will be demonstrated in the following exercise.

EXERCISE: CREATING A NEW TOOLBAR

1. Repeat steps 1 and 2 of the preceding exercise, Adding a Tool to a Toolbar.

2. As you did in step 3 in the preceding exercise, find the first tool/command icon that you want to include in your new toolbar.

3. Drag and drop the tool icon into the drawing area. AutoCAD makes a new toolbar with your icon choice as its only tool. This toolbar is assigned the name Toolbar1.

4. Repeat steps 2 and 3, adding as many tools as you want. The toolbar shown in Figure 5–10 consists of four dimension tools and the Erase tool.

5. To rename the toolbar, in the Customize dialog box click on the Toolbars tab to make the Toolbars folder active.

6. In the Toolbars window, scroll to find the Toolbar1 item and select it.

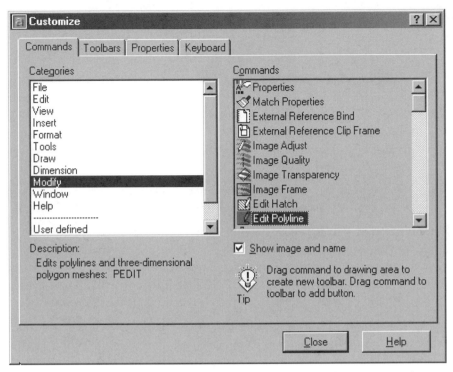

Figure 5–7 *A library of Modify commands is shown in the Commands window.*

Figure 5–8 *The PEDIT command and icon added to the Modify toolbar.*

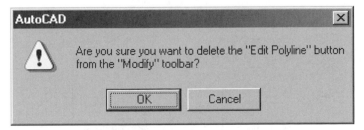

Figure 5–9 *AutoCAD issues a warning before allowing a toolbar icon to be deleted.*

Figure 5–10 *Creating a new toolbar with tools of your choice.*

7. Select the Rename button and, in the Rename Toolbar dialog box, enter a new name for the toolbar and select OK (see Figure 5–11). AutoCAD renames the toolbar and adds it alphabetically to the list of toolbars in the Toolbars window, as shown in Figure 5–12.

8. Remove AutoCAD from customize mode by selecting Close on the Customize dialog box. Any changes you made are written to the menu's MNS file.

The second method of creating a new toolbar is similar to the first but involves the use of the New button on the Toolbars folder of the Customize dialog box. Selecting this button causes AutoCAD to create an empty toolbar similar to that shown in Figure 5–13 after you enter a name for the toolbar in the New Toolbar dialog box. You then add tools to the new toolbar as you did in the preceding exercise

Modifying existing toolbars and creating new customized toolbars are relatively straightforward procedures that require no special knowledge of AutoCAD's menu macro language and can significantly increase your efficiency.

ADDING AN ITEM TO THE CURSOR MENU

Modifying AutoCAD's pull-down menus is a common and productive means of customizing AutoCAD. Most modifications of this type, however, require knowledge of AutoCAD's menu macro language, which is beyond the scope of this chapter. On the other hand, some simple additions to AutoCAD's cursor menu, for example, can be easily accomplished and can be real productivity boosters..

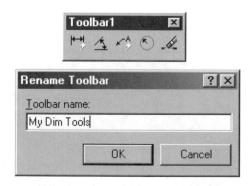

Figure 5–11 *Renaming a new toolbar.*

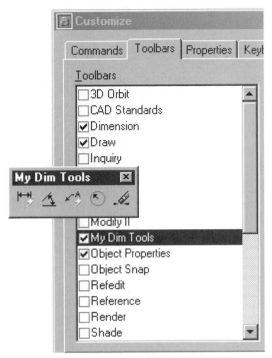

Figure 5–12 *The renamed toolbar.*

Figure 5–13 *An empty toolbar created by using the New button on the Toolbars folder.*

 Note: With AutoCAD in customize mode you can also copy icons from any visible toolbar to any other visible toolbar by holding down the Ctrl key while dragging and dropping the icon.

The cursor menu, also often referred to as the Osnap menu, is a popular target for modification because it is always close at hand. If the MBUTTONPAN system variable is set to zero, you can display it by clicking the system pointing device's middle button or wheel; otherwise, you display it with a Shift key/right-click combination. Figure 5–14 shows AutoCAD 2002's default cursor menu.

 Note: Before performing any modification to any AutoCAD menu file, make a backup copy and place it in a separate folder. You may want to consider creating a dedicated folder in which to store copies of all the AutoCAD files that you use in customization.

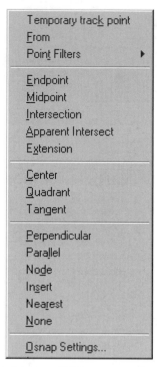

Figure 5–14 *AutoCAD's cursor or Osnap menu.*

To perform any AutoCAD pull-down menu customization, you must first find and open the main AutoCAD menu source file. The menu source file is named ACAD.MNS. In a standard AutoCAD installation, it is placed in the AutoCAD\ folder. As explained earlier in this chapter, this is a plain-text ASCII file that should be opened in a generic text editor program such as Windows Notepad or Wordpad.

 Note: During standard AutoCAD installation, Windows Notepad is automatically associated with the ACAD.MNS file. Double-clicking on this file in Windows Explorer will cause it to be opened in Notepad or in another file editor application that you may have designated during installation. The ACAD.MNS file is large and may not be accommodated by Windows Notepad. In this case, Windows will offer to open it in Wordpad, which is an acceptable alternative.

Once you have the ACAD.MNS file opened in your text editor, find or scroll down to the ***POP0 section of the file. You will find it at approximately line number 42 of the file. The beginning of this menu section appears as follows:

```
***POP0

**SNAP

          [&Object Snap Cursor Menu]
```

```
ID_Tracking      [Temporary trac&k point]_tt
ID_From          [&From]_from
ID_MnPointFi     [->Poin&t Filters]
ID_PointFilx      [.X].X
ID_PointFily      [.Y].Y
ID_PointFilz      [.Z].Z
                  [--]
ID_PointFixy      [.XY].XY
ID_PointFixz      [.XZ].XZ
ID_PointFiyz      [<-.YZ].YZ
                  [--]
```

The POP0 section is denoted by the ***POP0 line. The next line, **SNAP, denotes the name of the menu itself. The next line is not used by the Osnap menu and can be disregarded. Then follow the name tags, labels, and command sequences for all the entries on the Osnap menu. To the left are the name tags, all beginning with "ID_". These are unique tags for each line containing a function, and they are used for cross-referencing and for defining the Help strings that appear at the bottom of the screen on the mode status line. For our purposes in this chapter, these name tags can be ignored. Once you carry out the modification to this file, as you will do in the exercise that follows, you will save the changes to this file.

In addition to the name tag, each operative line in the POP0 section contains a label and a command sequence. For example, the first operative line has the name tag ID_Tracking followed by the label "Temporary trac&k point." The label is what actually appears on the displayed menu, as shown in Figure 5–14. The label is enclosed in square brackets. The ampersand symbol (&) that appears in the label is used to designate the shortcut letter for the line and can be disregarded for the Snap menu. The command sequence follows the label. This sequence is encoded in menu macro language and represents the commands that are executed when you select the item on the menu with the mouse. In this case, the command sequence is _tt, which initiates a temporary tracking point during commands.

In the following exercise you will edit the ACAD.MNS file and add the ability to initiate a transparent call to the CURSORSIZE system variable, allowing you to change the size of AutoCAD's cursor on the fly.

EXERCISE: ADDING A COMMAND TO THE CURSOR MENU

1. With AutoCAD not running and the ACAD.MNS menu file loaded in a text editor, find the beginning of the ***POP0 section.

2. Using the text editor, add the middle line of the following, shown in bold for clarity. (Do not forget the apostrophe preceding the word "cursorsize.")

```
                   [&Object Snap Cursor Menu]

ID_Csize           [Change Cursor Si&ze]'cursorsize

ID_Tracking        [Temporary trac&k point]_tt
```

3. Save and close the file.

4. Start AutoCAD. (Because the ACAD.MNS file has been altered, AutoCAD will compile a new ACAD.MNC file, as explained earlier in this chapter.)

5. Start the LINE command. At the "Specify next point or [Undo]" prompt, display the cursor menu (Shift+right-click, or click the middle mouse button). The cursor menu should appear as shown in Figure 5–15.

6. Select Change Cursor Size. At the "CURSORSIZE <5>" prompt, type **50** and press Enter. The cursor should change to 50 percent of the screen height.

7. Press ESC to abort the LINE command.

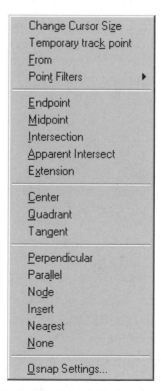

Figure 5–15 *The modified cursor menu.*

Although AutoCAD's menu macro language is not the subject of this chapter's survey of simple methods of customizing AutoCAD, you can see that by studying the ACAD.MNS file along with the techniques used in this section and the preceding exercise, you can easily modify the cursor menu with little effort.

SUMMARY

There are many ways to customize AutoCAD. Adding or changing command aliases in the PGP file, modifying or creating toolbars, and modifying menus are among the most popular. As this chapter demonstrated, these methods do not require any advanced knowledge of programming techniques, and they are easy to learn.

CHAPTER 6

Advanced Customization

In the preceding chapter you learned how to carry out basic AutoCAD customization without having to be proficient in any programming language. In this chapter you'll see that investing the time and effort to learn about the structure of AutoCAD's various menu files and the relatively simple "language" used in them will allow you to advance even further in the customization of AutoCAD.

AutoCAD is the most popular CAD software for use on PC desktops and workstations in the world today. The nearly 4,000,000 AutoCAD users are found in some eighty countries around the world and have made AutoCAD the defacto standard for producing computer-generated drawings. One of the primary reasons AutoCAD has gained this leadership position is its so-called "open architecture," which means its ability to be programmed, or customized, by its users and independent third-party software developers. As an AutoCAD user, for example, you can easily modify the AutoCAD interface to suit your particular needs, and you can even add new commands. Third-party developers can write and integrate complex applications that perform specialized tasks within the basic AutoCAD framework. Customization allows AutoCAD to fit the needs of its users, increasing their efficiency.

As important as the ability to customize AutoCAD is the ease with which much of the customization can be carried out. The average user, once he or she gains a moderate degree of proficiency with AutoCAD, can quickly learn how to modify the basic AutoCAD menu structure—or even write specialized menus—and format new toolbars to increase their AutoCAD working efficiency.

In this chapter you will learn how to improve your productivity by formatting customized AutoCAD menus. You will learn how to load different menus, build a customized menu toolbar, and use menu macros to automate your work environment

This chapter will cover the following topics:

- Understanding AutoCAD menu files

- Understanding menu file sections

- Using the ***Buttons and ***Aux sections

- Using the ***POP*n* sections

- Creating menu macros

- Building menu macros

- Using AutoLISP in menu macros

- Adding a customized tool to a toolbar

CUSTOMIZING MENUS

The basic menu structure of AutoCAD provides you with an effective method of executing nearly every AutoCAD command. This basic menu is necessarily designed to meet the needs of most AutoCAD users, most of the time. However, by customizing the basic menu or by writing and loading new custom menus, you can add new command functions, gather the commands you use most often together in a convenient place, or add menu macros that perform repetitive or specialized tasks with less effort. If you have custom AutoLISP functions, they can also be called from menus. You can also create customized menus for task-specific functions. AutoCAD menus, including ones you create, are portable so that you can take them with you to different locations.

UNDERSTANDING AUTOCAD MENU FILES

The menus used by AutoCAD are defined by various menu files. The files that you will be concerned with in customization work are in standard text file format, which you can easily modify. You can also define new menus. When you create or edit a menu, you assign items to the menu. You associate with these items menu macros that perform specific AutoCAD functions. A menu macro can be as simple as a sequence of standard AutoCAD commands, or as complex as a combination of commands and AutoLISP or DIESEL code or a combination of all three. Menu macros can be thought of as "command strings." When the user picks a menu item, the associated macro is executed. You will learn how to construct menu macros later in this chapter. The menu files also describe the appearance and position of menu items in relation to the overall AutoCAD user interface. There are several types of menu files, each distinguished by its file extension. These types include MNU, MNS, MNC, MNR, and MNL files. The base AutoCAD menu, for example, is called ACAD.MNU. For this base file there will also be an associated ACAD.MNS, ACAD.MNC, ACAD.MNR, and ACAD.MNL file. The function and format of the various menu file types are described in Table 6–1.

When you start AutoCAD, the last menu used is automatically loaded. The name of this menu is stored in the system registry. You can manually load a different menu using AutoCAD's MENU or MENULOAD command. Whether loaded automatically or manually, AutoCAD finds and loads the specified menu using the following search sequence:

1. AutoCAD looks for an MNS source file of the specified name, following the AutoCAD Library Search Path.

   - If an MNS file is found, AutoCAD looks for the compiled (MNC) version of the same file in the same directory. If AutoCAD finds an MNC file with the same or later date and time as the MNS file, it loads this MNC file. Otherwise, AutoCAD compiles the MNS file, generating a new MNC file, and then loads the compiled MNC file.

   - If an MNS file is not found, AutoCAD looks for a compiled (MNC) menu file of the specified name in the Library Search Path. If AutoCAD finds this MNC file, it loads it.

   - If AutoCAD finds neither an MNS or MNC file, it searches the Library Search Path for a menu template (MNU) file of the specified name. If found, the MNU file is used to generate an MNS source file. This MNS file then generates a compiled MNC file, and AutoCAD loads this MNC file.

   - If AutoCAD finds no menu files of the specified name, it generates an error message prompting you for another menu name.

2. After finding (or compiling) and loading the MNC file, AutoCAD searches the Library Search Path for a menu LISP (MNL) file. If AutoCAD finds this file, it evaluates the LISP code and loads it into memory.

3. Anytime AutoCAD compiles an MNC file it also generates a resource (MNR) file containing the toolbar icon bitmap definitions used by the associated menu.

As you can see by looking at Table 6–1 and examining the menu loading procedure, it is the compiled MNC version of any given menu that is actually loaded. Also note that the MNC file is compiled from the MNS file, not the MNU file. In fact, the MNU template file is essentially useless in a Windows environment, and, as you will see when we discuss custom toolbars later in this chapter, the MNU file can be dangerous. It is therefore recommended that you move the MNU file (e.g., ACAD.MNU) out of the Library Search Path so that AutoCAD never reads it

Table 6–1: *Menu File Types*

Menu Type	Description
MNU	Template menu file in standard (ASCII) text format
MNS	Menu source file in standard (ASCII) text format; generated by AutoCAD from the **MNU** file
MNC	Binary format file compiled from the **MNS** file; the file that is actually loaded
MNR	Binary file containing the bitmaps used by the menu
MNL	Menu file in standard (ASCII) text format containing AutoLISP code required by the menu

Note: The Library Search Path consists of the Support Files Search Path, which is specified under Support Files Search Path on the Files folder of the OPTIONS command's dialog box, and the following locations: the current directory (typically determined by the "Start In" setting in your shortcut icon), the directory that contains the current drawing file, and the directory that contains the AutoCAD program files. Some of these search locations may overlap.

when menu files are loaded or reloaded. *Perform all of your menu customization using the menu's MNS file.*

Tip: There are several ways you can keep the MNU file from interfering with the MNS and MNC files. You might leave the MNU file in its original location and rename it as ACAD-TEMPLATE.MNU, for example. Or you could create a new directory named Safe, for example, and move the ACAD.MNU file there. The second method has the advantage of providing a directory in which to store other AutoCAD files that you wish to protect—such as the original ACAD.PGP or ACAD.LIN files. Never delete the original MNU file. In an emergency, it *can be used to rebuild the MNS file, although you may lose much of your menu customization.*

UNDERSTANDING MENU FILE SECTIONS

All AutoCAD menu files are made up of one or more major sections. Each section is associated with a specific area of the AutoCAD menu interface, such as the main (or pull-down) menu, the toolbars, etc. Each menu section contains menu entries that provide instructions for the appearance and action associated with the menu entry. Each menu section is identified by a section label having the form ***section name. The various section labels and their associated menu areas are listed in Table 6–2.

Table 6–2: *Menu Section Labels*

Section Label	Menu Area
***MENUGROUP	Menu file group name
***BUTTONS*n*	Pointing-device button menu
***AUX*n*	System pointing-device menu
***POP*n*	Pull-down/shortcut menu areas
***TOOLBARS	Toolbar definitions
***IMAGE	Image-tile menu area
***SCREEN	Screen menu area
***TABLET*n*	Tablet menu area
***HELPSTRINGS	Text displayed in the status bar when a pull-down or shortcut menu item is highlighted or when the cursor is rested on a toolbar item
***ACCELERATORS	Accelerator key definitions

Within a given menu section there can be one or more *alias* sections of the form **alias. Note that menu section labels have the three-asterisk (***) prefix, while alias labels have a two-asterisk (**) prefix. For the customization done in this chapter, you can ignore alias labels.

You can include comments within a menu file for documentation or notes by placing the comments on a line that begins with two forward slashes (//). These lines are invisible to the AutoCAD menu interpreter. Such notes are often useful when you view the file in a text editor. There are several useful and informative comment lines found in the ACAD.MNU file, for example.

The various sections of an AutoCAD menu differ in their functionality and appearance. Descriptions of the ***BUTTONS, ***AUX, and ***POP sections follow in this chapter. Several of the menu types are shown in Figure 6–1. For this chapter we will limit our menu file customization to the ***POP sections. Later in the chapter we will use the interface features of AutoCAD 2002 to customize toolbars. Macros for ***POP sections are also considered later in this chapter.

USING THE ***BUTTONS AND ***AUX SECTIONS

The Buttons and Aux sections appear near the top of the standard ACAD.MNU and ACAD.MNS files and are used to customize your system pointing device. The two sections are functionally identical, and which of the two your system uses depends

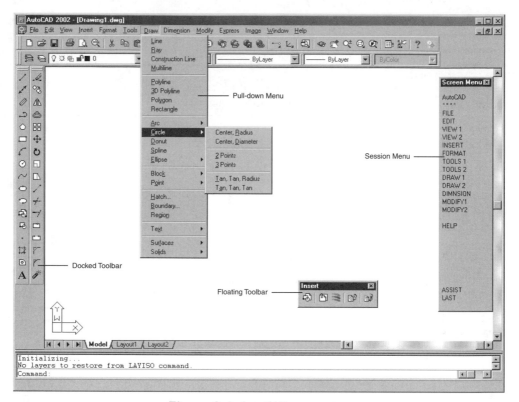

Figure 6–1 *AutoCAD menu types.*

upon the type of pointing device you use. The standard Windows system mouse uses the Aux section and any other input device (a digitizer puck, for example) uses the Buttons section. In this chapter we will confine our discussion to the Aux (mouse button) section.

All pointing devices have a pick button, which is used to specify points and select objects on the screen. On a normally configured, standard Windows mouse, this is the left button. On most digitizing pucks, this button is normally labeled button 1. This pick button is not customizable, and no accommodation for its function is found in AutoCAD menus. The remaining button(s) can have commands, functions, or macros assigned to them.

The AUX*n* sections of the menu file define the actions (commands or macros) associated with the buttons on your mouse. Each line in any given section (AUX1 or AUX2, for example) represents a mouse button. You can access each button menu with the key/button combination shown in Table 6–3.

Table 6-3: *Buttons and Associated Menu Sections*

Key/Button Combination	Menu Section
Simple button press	AUX1 and BUTTONS1
Shift + button press	AUX2 and BUTTONS2
Ctrl + button press	AUX3 and BUTTONS3
Ctrl + Shift + button press	AUX4 and BUTTONS4

The first lines after menu section label ***AUX1 or ***BUTTONS1 as they appear in the ACAD.MNU file are used only when the SHORTCUTMENU system variable is set to zero. If SHORTCUTMENU is set to a value other than zero, the built-in menu is used. Similarly, the second lines after the ***AUX1 or ***BUTTONS1 labels are used only when the MBUTTONPAN system variable is set to zero.

Consider the following ***AUX1 section as it appears in ACAD.MNS. The items in square brackets do not appear in the actual menu; they were only added here for clarity. (Refer to Table 6-4 for an explanation of menu macro codes.)

```
***AUX1
;   [button no.2 Enter]
^C^C[button no.3 Cancel]
^B  [button n0.4 Toggles Snap]
^O  [button no.5 Toggles Ortho]
^G  [button no.6 Toggles Grid]
^D  [button no.7 Toggles Coords]
^E  [button no.8 Cycles to next isometric plane]
^T  [button no.9 Toggles Tablet]
```

Remember that the pick button cannot be assigned and is therefore not included in the menu. The first line after the ***AUX1 section label (;) therefore represents the next button after the pick button—usually button 2 or the right mouse button. On a two-button mouse, therefore, only the first line of this listing has any meaning. The second line (^C^C) after the section label applies to button three on a three-button mouse, and the remaining lines having no meaning. Later you will see that the semicolon (;) and the ^C^C codes represent an Enter and a Cancel command, respectively.

Now consider the ***AUX2 section:

```
***AUX2
$P0=SNAP $p0=* [button no.2]
$P0=SNAP $p0=*[button no.3]
```

Referring to Table 6–3, you can see that the ***AUX2 section defines the actions taken when a combination of Shift and another button is pressed. In this ***AUX2 listing there is a defined action for Shift + button 2 and Shift + button3. Again, with a two-button mouse, the second code line has no meaning. The two lines in this example are identical, and the code causes the ***POP0 cursor menu to be loaded and displayed using the following menu code format:

```
$Pn=name $p0=*
```

Here, $ is the special character code for loading a menu area; pn specifies the POP menu section, with n being the number of the section. The =* symbols force a display of the named section. You will learn more about POP menu sections later in this chapter. It is the POP0 section and code that displays the POP0 cursor menu when you Shift + right-click.

The ***AUX and ***BUTTONS sections of the ACAD.MNS provide several opportunities for customization of the functions performed by your input device. If you use a ten- or twelve-button digitizer puck, you can program up to twenty-seven or thirty-three button functions, respectively. Even with the standard two-button mouse, the right button can be programmed to perform three different tasks.

USING THE ***POP*n* SECTIONS

The ***POP*n* menu sections contain the definitions for the pull-down and shortcut menus. These menus are displayed in a "drop-down" or cascading fashion, as shown in Figures 6–2 and 6–3. Pull-downs have the advantage of being easy to read and use, while only temporarily taking up screen space.

Shortcut menus are usually displayed while commands are in progress, and they often duplicate the various command options available and displayed at the command line. Shortcut menus are displayed at or near the crosshairs or cursor in the screen area. A typical shortcut menu (for the PEDIT command) is shown in Figure 6–3.

Pull-down menus are defined in the ***POP1 through ***POP499 menu sections. Shortcut menus are defined in the ***POP0 and ***POP500 through ***POP999 sections. AutoCAD constructs the menu bar across the top of the screen for the ***POP1 through ***POP16 menu sections. If no ***POP1 through ***POP16 definitions are found, AutoCAD constructs a menu bar containing a default File and Edit menu only.

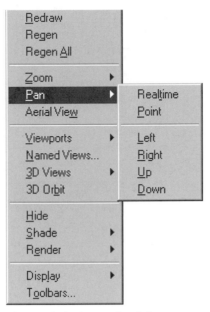

Figure 6–2 *A typical pull-down menu.*

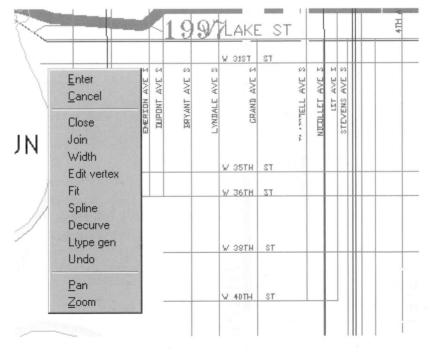

Figure 6–3 *A typical shortcut menu.*

The following example shows a typical pull-down menu. The actual appearance of the Format menu described is shown in Figure 6–4. The specific syntax and menu code for pull-down menus will be discussed later in this chapter.

```
***POP5

**FORMAT

ID_MnFormat      [F&ormat]

ID_Layer         [&Layer...]'_layer

ID_Ddcolor       [&Color...]'_color

ID_Linetype      [Li&netype...]'_linetype

ID_Linewt        [Line&weight...]'_lweight

                 [--]

ID_Style         [Text &Style...]'_style

ID_Ddim          [&Dimension Style...]'_dimstyle

ID_PlotStyle     [$(if,$(eq,$(getvar,pstylemode),1),~,)Plot
   St&yle...]^C^C_plotstyle

ID_Ddptype       [&Point Style...]'_ddptype

ID_Mlstyle       [&Multiline Style...]^C^C_mlstyle

                 [--]

ID_Units         [&Units...]'_units

ID_Thickness     [&Thickness]'_thickness

ID_Limits        [Dr&awing Limits]'_limits

                 [--]

ID_Ddrename      [&Rename...]^C^C_rename
```

In this example, each functional line of the menu consists of a name tag, a label, and a menu macro. Menu macros are discussed in the following section.

CREATING MENU MACROS

Now that you have a basic understanding of the ***Aux, ***Buttons, and ***Pop sections of an AutoCAD menu, it is time to learn how to build new menu items and place them into a menu. To do this you will need to know the special codes used in menu macros. If you want to include command parameters in a menu item, you will also need to know the sequence in which the command expects its parameters. Every character in a menu macro is significant, even blank spaces. We will devote our attention to the ***Pop menu section.

First of all, you should know an important definition: A menu macro is a shorthand or coded method of representing keystrokes that you would type at the command prompt. When AutoCAD reads the macro code, in effect it duplicates the keystrokes just as though they had been typed at the keyboard. There is a group of

Figure 6–4 *The Format pull-down menu.*

codes used to represent certain keyboard entries. These special characters are listed in Table 6–4.

Although this code listing may seem daunting, several of the characters are rarely used. Perhaps five or six are used frequently. You may need to refer to Table 6–4 frequently until you become familiar with the more commonly used codes.

The general form of functional menu entries is consistent in the ***POP*n* sections. The general format is as follows:

```
Name_tag        [label]menu_macro
```

The following line, taken from the Modify menu, is typical:

```
ID_Pedit        [&Polyline]^C^C_pedit
```

At least one space should separate the Name_tag element from the label element. There can be no space between the label element and the menu_macro element. The label must be enclosed within a set of square brackets. In the example, the first item, ID_Pedit, is the name tag for the entry. The label, [&Polyline], displays the word polyline on the pull-down. The ampersand character (&) appears immediately before the alphanumeric character in the label that appears underscored on-screen, indicating that the character is the Alt+character hot key for the entry. The menu macro in this example consists of the code ^C^C_pedit, which issues a cancel followed by AutoCAD's PEDIT command. When the Line item is selected from the standard Modify pull-down menu, this is the menu macro that is executed.

Table 6–4: *Special Characters Used in Menu Macros*

Character	Description
;	Issues Enter
^M	Issues Enter
^I	Issues Tab
SPACEBAR	Enters a space; a blank space between command items in a menu entry is equivalent to pressing the Spacebar
\	Pauses for user input
.	A period can be used immediately prior to a native AutoCAD command to override any redefinition of that command that may be in effect.
_	An underscore character; translates AutoCAD commands and keywords that immediately follow
+	Continues menu macro to the next line if last character on a line
=*	Displays the current top-level image, pull-down, or shortcut menu
*^C^C	Prefix for a repeating item
$	Loads a menu section or introduces a condition DIESEL macro expression ($M=)
^B	Toggles Snap on or off (Ctrl+B)
^C	Cancels command (Esc)
^D	Toggles Coords on or off (Ctrl+D)
^E	Sets the next isometric plane (Ctrl+E)
^G	Toggles Grid on or off (Ctrl+G)
^H	Issues backspace
^O	Toggles Ortho on or off (Ctrl+O)
^P	Toggles MENUECHO on or off
^Q	Echoes all prompts, status listings, and input to the printer (Ctrl+Q)
^T	Toggles Tablet on or off (Ctrl+T)
^V	Changes current viewport (Ctrl+V)
^Z	Null character that suppresses the automatic addition of SPACEBAR at the end of a menu item

Note: Name_tag elements are used as "aliases" or a shorthand method of referring to the associated menu macro from other portions of the menu, most notably from the Accelerator key definitions. They also provide the link to the ***Helpstrings menu section. As such, they are beyond the scope of this chapter, although they will be included in the chapter's examples.

BUILDING MENU MACROS

In this section you will actually construct several menu macros and place them in a separate pull-down menu that you will incorporate into your current main menu.

For the purpose of gaining practice constructing menu macros, we will assume that your current menu is defined in the file ACAD.MNS. You should review the various menu file types discussed earlier in this chapter. The distinction between ACAD.MNU and ACAD.MNS should be especially kept in mind. If your current menu is other than ACAD.MNS, substitute your menu name in the examples that follow or load ACAD.MNS using the MENU command.

Note: As you work with the menu macros in the exercises that follow, you will modify the base menu. It is important that you make a backup copy of this menu so that you can restore a known, working menu if necessary.

Once you have a safe copy of the ACAD.MNS file, you can proceed to make your own pull-down menu. Recall that the pull-down menus are called ***POP*n* menus, where *n* is a number from one through 499. As it comes out of the box, AutoCAD has eleven menu-bar pull-down menus—POP1 through POP11. In the following exercise you will add a POP12 pull-down menu named MyMenu. This pull-down menu will contain several types of menu macros, which you will add in the following exercises. (The new POP12 menu will initially appear to the right of the Help pull-down menu. You will later move it to between the current Modify and Window pull-downs.)

EXERCISE: ADDING A NEW PULL-DOWN MENU

1. With AutoCAD running, load the ACAD.MNS file into a text editor.

2. Scroll down or use the text editor's Find function to navigate to the ***POP500 section of the menu file. Immediately above the ***POP500 line, add the following new lines:

```
***POP16
**MYMENU
ID_Mymenu [&MyMenu]
ID_Break_f[&Break-F]^C^C_Break;\f;\
```

3. Check your typing closely. The number of spaces between the Name_tag and label elements is not important. Format these elements so that they are easy to read. Use the menu text that appears in the ***POP500 section immediately below this section as a guide.

4. Save the changes you have made to the file and minimize the text editor.

5. In AutoCAD, from the Tools menu, choose Customize>Menus to display the Menu Customization dialog box. Make sure the Menu Groups folder is active, as shown in Figure 6–5.

6. If necessary, choose the ACAD menu from the Menu Groups window and then choose Unload to unload the menu.

7. Next choose Browse and, in the Select Menu File dialog box, make sure that the Files of Type window is set to show MNC and MNS file types (see Figure 6–6).

8. In the Select Menu File dialog box, find and select the ACAD.MNS file as shown in Figure 6–6. Choose Open.

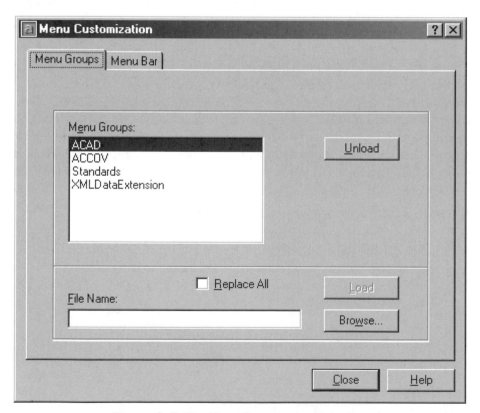

Figure 6–5 *The Menu Customization dialog box.*

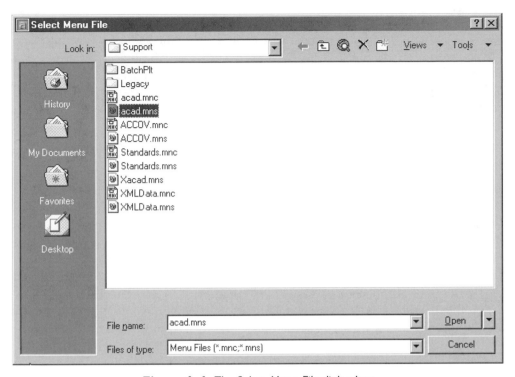

Figure 6–6 *The Select Menu File dialog box.*

9. In the Menu Customization dialog box, make sure that the ACAD.MNS file is shown in the File Name window and then choose Load. AutoCAD uses the new MNS file to generate a new MNC file and then loads it. The new MyMenu pull-down menu should appear (grayed-out) to the right of the Help pull-down menu.

10. To reposition the MyMenu pull-down menu, choose the Menu Bar tab to make the folder active.

11. On the Menu Bar folder, under the Menus window, select the MyMenu item.

12. In the Menu Bar window, select the Windows item and choose the Insert>> button. Note that the MyMenu item is repositioned to the left of the Window item. Choose the Close button to close the Menu Customization dialog box.

13. Select the MyMenu pull-down menu. The menu should resemble Figure 6–7.

Figure 6–7 *The new MyMenu pull-down menu.*

Dissecting the New Break-F Macro

AutoCAD's native BREAK command has often been the subject of menu macro customization. In the BREAK command, if you select the object by using your pointing device, AutoCAD both selects the object and treats the selection point as the first break point. At the next prompt you can continue by specifying the second point or by overriding the first point by typing an "f." In many cases, this default mechanism of having the object pick point become the first specified break point is inconvenient. The following menu macro automatically enters the overriding "f":

```
[&Break-F]^C^C_Break;\f;\
```

For an explanation of this macro, refer to Table 6–4.

Break-F is a typical menu macro. It uses only standard AutoCAD command input and represents nothing more than a means to supply automatically the equivalent of conventional keyboard input. In constructing a menu macro such as Break-F, it is useful to run the command sequence "manually," noting the exact keyboard input you supply in response to each issued AutoCAD prompt. It is often helpful to write out the prompt and keyboard response sequence before attempting to duplicate the procedure in macro code. It is easy to forget, for example, that in the place of every keyboard Enter, you must use either a semicolon or space in the macro, or that the backslash character is used to represent either screen picks or typed responses to prompts.

Tip: Reviewing the command and response sequence for macros you wish to construct is often easier if you carry out the procedure and then use AutoCAD's text window to visually review the sequence as it took place at the command prompt. Recall that the F2 key displays this text screen.

Like many menu macros, the Break-F macro is quite simple. It gains its effectiveness in the time and effort it saves. In the next section you will see how AutoLISP can be utilized in another simple but effective menu macro.

USING AUTOLISP IN MENU MACROS

You can use AutoLISP expressions and variables to create menu macros ranging from very simple to extremely complex. One of the principal advantages of AutoLISP programming is the ability to "branch" into two or more programming directions, depending upon user input or the state of the drawing environment. This can lead to rather complex macros that possess "intelligence" and the ability to make "decisions." Shorter, less complex AutoLISP macros can also exhibit this decision-making ability as well. Often these less-complex macros utilize the state of AutoCAD's system variable settings to perform useful tasks that would be more cumbersome and time-consuming to perform from the keyboard or through a series of toolbar or menu picks.

The system variable CURSORSIZE records the current size of the screen cursor or crosshairs as a percentage of the screen size. Valid settings range from 1 to 100 percent. When set to 100, the crosshairs are full-screen and the ends of the crosshairs are never visible. When less than 100, the ends of the crosshairs may be visible when the cursor is moved to one edge of the screen. Some users prefer a small CURSORSIZE in certain situations (such as when working with 3D objects or with views other than plan view), while they prefer a full CURSORSIZE when working in plan views or while editing 2D objects.

In the following menu macro, AutoLISP functions are used to obtain the value of the CURSORSIZE system variable. If the current value is 100, an AutoLISP function is used to set the variable to a value of 5. If the variable has a current value other than 100, it is set to 100. Utilizing this macro, it is easy to flip the size of the cursor back and forth between its smallest and largest size. The macro is as follows:

```
^C^C(setvar "cursorsize" (if (= 100 (getvar "cursorsize")) 5 100))
```

It is interesting to realize that this macro is a combination of menu macro code (as outlined in Table 6–4) and AutoLISP code. The ^C^C is standard menu macro code, while the remainder of the macro is expressed in AutoLISP code. The two are compatible and work together effectively. Refer to Chapter 8, Introduction to AutoLISP, for an explanation of basic AutoLISP functions.

In the following exercise, you will add the CURSORSIZE menu macro to your MyMenu pull-down menu.

EXERCISE: ADDING A MACRO TO A PULL-DOWN MENU

1. With AutoCAD running, load the ACAD.MNS menu file into a text editor as you did in the preceding exercise.

2. Find the ***POP12, MyMenu pull-down menu that you added before. Add the following line:

```
ID_CursorF      [&Cursor-flip]^C^C(setvar "cursorsize" (if (=
   100 (getvar "cursorsize")) 5  100))
```

3. Check your typing and save the file. Minimize the text editor.

4. Repeat steps 5-12 of the preceding exercise to load the new ACAD.MNS file and reposition the menu.

   The MyMenu pull-down menu should resemble Figure 6–8.

5. Test the Cursor Flip macro to verify that the cursor flips back and forth between the two sizes.

ADDING A CUSTOMIZED TOOL TO A TOOLBAR

In an earlier section of this chapter (Building Menu Macros), we developed a menu macro that called the BREAK command and automatically supplied the "f" for "specify first point" option for the command. This allowed you to select the object to be broken (line, arc, etc.) and then specify the first break point. You may want to review that section to reacquaint yourself with the basic macro mechanism. The macro code, without the name tag or menu label, is as follows:

Figure 6–8 *The new MyMenu pull-down menu.*

```
^C^C_break;\f;\
```

In the following exercise we will utilize this same macro and incorporate it into a new tool that we will add to the Modify toolbar. We will also "design" a new icon to accompany the tool.

EXERCISE: ADDING A CUSTOMIZED TOOL TO A TOOLBAR

1. With AutoCAD running, make sure that the Modify Toolbar is displayed. Then from the Tools menu, choose Customize>Toolbars.

2. In the Customize dialog box, choose the Commands tab to make the folder active.

3. Near the bottom of the Categories window, select User Defined.

4. In the Commands window make sure that the User Defined button is selected and drag it to the Modify toolbar, positioning the horizontal "I-beam" symbol above (or to the left of) the icon for the standard BREAK command. Release the left mouse button. Note that an empty icon space is created (see Figure 6–9).

Figure 6–9 *Adding a blank icon to a toolbar.*

5. Click in the empty icon space. The outline of the empty icon appears and the Button Properties folder of the Customize dialog box becomes active, as shown in Figure 6–10.

The Button Properties folder provides the means to enter the macro code for the new button. It also has facilities for selecting an icon from AutoCAD's library of icons and designing a new icon for the new tool. You can also designate the text for the buttons ToolTip and its help string. We will use the existing BREAK command's icon as a starting point for a new icon.

6. First, in the window containing the selection of predefined icons, select the icon for the BREAK command. AutoCAD displays an enlarged depiction of the icon on a Button Image tile (see Figure 6–11).

7. To begin customizing this icon, select the Edit button. The Button Editor dialog box is displayed as shown in Figure 6–12.

8. In the Button Editor, select Grid. Select the Pencil tool and the red color swatch. Carefully pencil-in the letter "F" as shown in Figure 6–13.

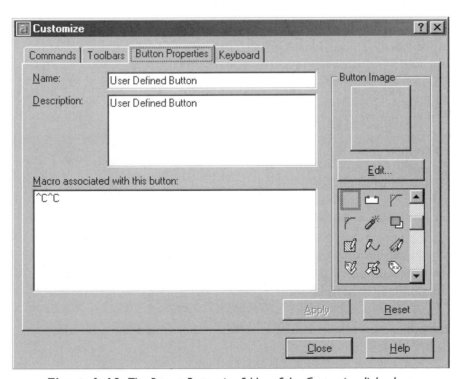

Figure 6–10 *The Button Properties folder of the Customize dialog box.*

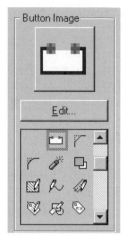

Figure 6–11 *Selecting the standard BREAK command's icon.*

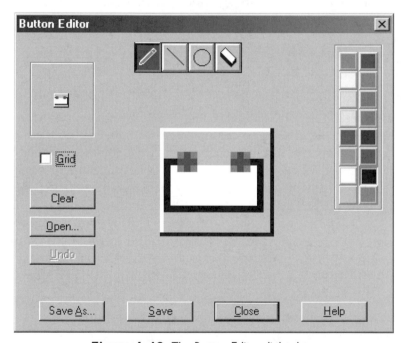

Figure 6–12 *The Button Editor dialog box.*

Working in the Button Editor is somewhat "tricky." If you make a mistake, you can cover it up by penciling over it in the background color and starting again. Use the preview tile in the upper-left corner of the editor to see how the button will look in its actual size.

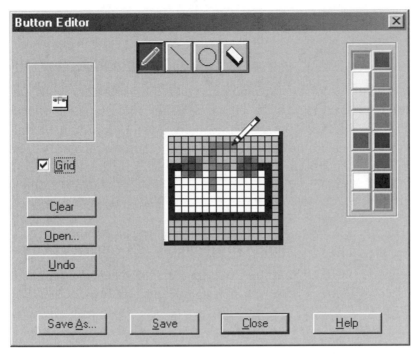

Figure 6–13 *Modifying an existing icon in the Button Editor.*

9. When you are satisfied with your icon design, select the Save As button and, in the Create File dialog box, create a BMP icon file in AutoCAD's \Support folder. Give the file a logical name such as Break-f.BMP. Close the Button Editor dialog box.

10. In the Customize dialog box, give the button a name in the Name input box and a description in the Description input box (see Figure 6–14).

11. Select the Apply button. The new icon appears on the Modify toolbar.

12. Close the Customize dialog box and test the new icon.

    If the macro fails to work as expected, the problem is most likely in the macro code. Check it carefully and correct it as necessary.

 Note: If you use the Edit function of the Button Properties dialog box to design your own button icons and then use the Save As button in the Button Editor to assign a name to the icon's BMP file, AutoCAD assigns a random name to the BMP file (for example, ICON4581.BMP) in the MNS file. This irksome behavior is a longstanding "bug" in the design of the Button Editor/Customize dialog boxes. Even in AutoCAD 2002, this bug persists and requires you to edit manually the MNS file's ***Toolbar section, replacing the random file name with the more meaningful name you assigned. In the MNS file, search for the toolbar name and manually edit the BMP file name.

Figure 6–14 *Assigning a macro, name, and description to a new button.*

In the following exercise you will edit the ACAD.MNS file's ***Toolbar section to replace the randomly generated BMP file name assigned by AutoCAD with the more meaningful name you assigned in step 9 of the preceding exercise. If you have not already done so, it is suggested that you make a backup copy of the current MNS file before performing this exercise.

EXERCISE: EDITING THE ***TOOLBARS SECTION OF THE MNS FILE

1. With AutoCAD *not* running, open the ACAD.MNS file in a text editor, as you have in the previous exercises in this chapter.

2. Using the text editor's Find facility, go to the ***Toolbars section. Narrow your search by finding the name you assigned to the button in step 10 of the preceding exercise. In this case, search for the text "break with first." The line(s) should appear something like this:

```
ID_UserButton_2 [_Button("Break with First", "ICON8215.bmp",
    ICON_16_BLANK")]^C^C_break;\f;\
```

3. Replace "ICON8215" (or whatever number has been assigned) with "Break-f.bmp" to make the line(s) read as follows:

```
ID_UserButton_2 [_Button("Break with First", "Break-f.bmp",
    ICON_16_BLANK")]^C^C_break;\f;\
```

4. Save the ACAD.MNS file, minimize the text editor, and start AutoCAD. The new Modify toolbar tool icon you designed in the preceding exercise should appear similar to that shown in Figure 6–15. When you are satisfied with your new tool, close the text editor.

Figure 6–15 *The new tool and icon on the Modify toolbar.*

SUMMARY

AutoCAD offers many different ways to customize both its commands and its command interface. You can even write "new" commands or modify commands to act in customized ways. Such customization offers the advantages of saving time and increasing your drawing efficiency. In this chapter you gained a basic understanding of AutoCAD menus and the specialized language used to build customized menu macros. You also learned how to modify standard toolbars and construct new ones. You now have the tools needed to move forward in customization.

SECTION

IV

**Developing
Custom
Applications**

Introduction to AutoLISP

CHAPTER 7

Some of AutoCAD's most powerful features are its programming interfaces. With support for AutoLISP, ARX, DIESEL, menu macros, and script programming, you can customize AutoCAD to just about any degree of sophistication you want.

This chapter focuses on AutoLISP—by far the most popular of AutoCAD's programming interfaces. The chapter will introduce you to the basics of AutoLISP and some of its capabilities. It is not intended to be an exhaustive survey of AutoLISP.

This chapter will cover the following topics:

- Introducing AutoLISP
- Using AutoLISP for keyboard macros
- Creating a simple AutoLISP routine
- Using ACAD.LSP and ACADDOC.LSP

INTRODUCING AUTOLISP

LISP is a programming language originally developed for use in the field of Artificial Intelligence (AI) in the early sixties. In fact, with the possible exception of FORTRAN, LISP is the only surviving high-level programming language from the sixties. There are some dozen dialects, or subsets, of LISP in use today. LISP is an acronym for LISt Processing, and all of LISP's data is, in fact, contained in the form of lists.

AutoLISP is Autodesk's own subset of the LISP language. Due to AutoCAD's popularity, there are today more worldwide users of AutoLISP than of any other implementation of this popular language. AutoLISP was introduced in an early version of AutoCAD and has remained a popular, although little changed, through the advent of Visual LISP as an add-on in AutoCAD Release 14. Some significant extensions to AutoLISP's capabilities and functionality became possible with the introduction of Visual LISP as an integral part of AutoCAD in AutoCAD 2000.

Programming with AutoLISP is easier than you might think. In fact, a large number of the many AutoCAD users who program with AutoLISP had no programming experience before learning AutoLISP. In this chapter you will be introduced to the basic concepts of AutoLISP programming.

The following is a sample of a simple AutoLISP macro:

```
(defun C:ZO ()
  (command "zoom" ".5X")
)
```

The next element of a standard AutoLISP macro is a call to a standard AutoCAD command. Logically enough, we use the AutoLISP function *command* to call AutoCAD commands. Once the command is called, or started, you must make provision to answer all the prompts associated with the command. In the case of C:ZO, the ZOOM command is started. The nine options to the ZOOM command are now available just as they would be at the command line. In the case of C:ZO, we want to issue a zoom with a Scale option. At the keyboard you would type **S** for Scale and then type the scale factor, or you would just enter the scale factor directly. Likewise in the C:ZO macro, we can directly submit a scale factor to the ZOOM command. C:ZO uses a scale factor of .5X, which will yield a zoom of one-half the current screen size, or a zoom-out of 100 percent. Once it is defined, running the C:ZO macro merely involves typing ZO followed by Enter.

It would be difficult to devise a macro simpler than C:ZO, yet such simple macros, once defined and loaded into memory, can, for repetitive command sequences, save significant amounts of time compared to typing the same sequence at the command line or even using a mouse to perform the same functions with picks from a displayed toolbar.

AUTOLISP VS. VISUAL LISP

A quick word about Visual LISP is needed here. Visual LISP itself will be discussed in Chapter 10, Introduction to Visual LISP. Suffice it to say that Visual LISP is an extension of AutoLISP. Ninety-nine percent of AutoLISP can be found incorpo-

Tip: The key to writing AutoLISP macros is knowing the AutoCAD commands and their options, as well as the type of information the various prompts and options require. You can usually save time and reduce the number of errors in your macros by first "running" them conventionally from the keyboard at the command line. You can then either write down the exact input and response sequence or change to the text screen (press F2) to view the sequence as you compose the macro.

rated into Visual LISP. Any time you invest into learning AutoLISP can be directly beneficial in working with Visual LISP. At the same time, Visual LISP improves upon AutoLISP in two significant ways. First, Visual LISP is embedded in a sophisticated Integrated Development Environment (IDE) that provides a number of features that make developing and debugging AutoLISP code and programs much easier. A built-in, AutoLISP-aware text editor, for example, makes checking and matching parentheses quick and easy. Secondly, Visual LISP incorporates a whole set of new ActiveX-like functions linking LISP to AutoCAD's object-based architecture. These "extensions" of AutoLISP make Visual LISP the first significant improvement or update to AutoLISP since its introduction early in AutoCAD's history. It must, however, have a good working knowledge of AutoLISP before you can effectively work in Visual LISP.

UNDERSTANDING AUTOLISP'S PARENTHESES

Before moving on to a closer look at AutoLISP keyboard macros, you need to understand the matter of LISP parentheses. We have said that the LISP programming language is based on lists, and that all data in LISP is stored and presented as lists. In LISP, the lists are delimited with parentheses; a single list is begun with a left, or *opening*, parenthesis and closed with a right, or *closing* parenthesis. Lists can—and usually do—contain other lists, each of which must, in turn, be delimited with opening and closing parentheses. In the macro C:ZO, for example, there are three lists:

```
(defun C:ZO ();line 1
    (command "zoom" ".5X")          ;line 2
)       ;line 3
```

The parent list holds the other two lists and begins on line 1 with the "(defun..." statement. The second list is an "empty" list; that is, it contains no data; it appears on line 1 also, as (), immediately after the name of the macro. The reason for the appearance of this empty list goes beyond the scope of this chapter; suffice it to say that it is required. The third list comprises line 2 of the code. It begins with the *command* function. Line three of the code contains only one item: the closing, or matching, parenthesis for the very first parenthesis on line 1. Although it is not required, for visual clarity it appears on a separate line and directly under the parenthesis it matches.

In the case of C:ZO, there are two lists within the parent, or outer, list. When a list appears as an element within a larger list, it is called a nested list. The nesting of lists to a depth of several levels is common in AutoLISP. Nesting, however, often makes the matching of all the right and left parentheses more difficult. If an AutoLISP function—whether a simple macro or a complicated multi-line routine—contains unbalanced parentheses, an error message will result when the function is run. Keeping track of parentheses is one of the less interesting aspects of programming in AutoLISP.

USING AUTOLISP IN KEYBOARD MACROS

Many of the most useful AutoLISP macros are as simple as the C:ZO you defined in the previous section. Other macros, while they perform more involved tasks, are extensions of the simple macro model represented by C:ZO. An AutoLISP macro, for example, may call a series of commands as shown later in this section, or it may start a command, provide predetermined responses to one or two command prompts, and then allow the user to finish the command sequence interactively.

The following is an example of an AutoLISP macro that calls more than one AutoCAD command:

```
(defun C:RX ()
  (command "UCS""X""90")
  (command "PLAN" "C")
)
```

This macro, C:RX, could be useful in 3D work. It first rotates the UCS about the X-axis by 90 degrees. It next calls the PLAN command and specifies the current plan view. When you type **RX** and press Enter, this macro performs the work of sixteen keyboard strokes.

Note: To help you better understand the logic and sequence of these macros, you may want to open a drawing and run the command involved in AutoCAD as you follow the explanations.

The next AutoLISP macro is an example of a macro that suspends command operation and allows the users to finish the sequence interactively:

```
(defun C:PB ()
(command "PURGE" "B" "*" "Y")
```

As you can see, this macro, C:PB, starts the PURGE command and answers the first prompt of the command with a "B" for Blocks. The next PURGE command prompt asks you to name the blocks to offer for purging or to enter an asterisk (*) for all blocks. The next prompt asks if you want to verify each name to be purged. In the C:PB macro, this prompt is answered with a "y" for "yes." At this point the macro ends and allows the PURGE command to continue normally, offering candidate blocks, if any, for purging. You can now accept or reject candidate blocks interactively as you would if the command had been entered manually.

Note: Many AutoCAD commands offer defaults for some of their prompts. Often these defaults are based on a previous response to the prompt. You can accept a default value to any prompt by supplying a pair of double quotation marks ("") in the macro—the equivalent to a keyboard ENTER. However, you should form the habit of explicitly specifying the response you wish for each prompt in your macros unless you are certain that the offered default is the response upon which you are basing your macros.

CREATING A SINGLE AUTOLISP FILE CONTAINING MULTIPLE MACROS

As you can see, AutoLISP macros are typically short, with perhaps only two or three lines of AutoLISP code. It is therefore frequently useful to group a number of related macros together in a single file that can be loaded in one operation. No matter how many individual macros are contained in a file, each is loaded into AutoLISP's memory when the file is loaded.

Note: AutoLISP code can be written with any standard ASCII text editor, such as Windows Notepad or WordPad. Third-party, ASCII text editor programs may also be used. The important thing to remember is to use a non-formatting ASCII editor. AutoCAD Releases 2000, 2000i, and 2002 also provide a full-featured, LISP-aware ASCII text editor as a part of Visual LISP. We will look at the basic elements of Visual LISP in Chapter 10, Introduction to Visual LISP.

A useful grouping of AutoLISP keyboard macros might have the manipulation of the UCS as its theme and include a family of macros related to the C:RX macro from earlier in this chapter. The following group of eight macros is just such a grouping:

```
(defun C:RX ()
  (command "UCS" "X" "90")
  (command "PLAN" "C")
  (princ)
)
(defun C:RX- ()
  (command "UCS" "X" "-90")
  (command "PLAN" "C")
  (princ)
)
(defun C:RY ()
  (command "UCS" "Y" "90")
```

```
    (command "PLAN" "C")
    (princ)
  )
(defun C:RY- ()
  (command "UCS" "Y" "-90")
  (command "PLAN" "C")
  (princ)
)
(defun C:RZ ()
  (command "UCS" "Z" "90")
  (command "PLAN" "C")
  (princ)
)
(defun C:RZ- ()
  (command "UCS" "X" "-90")
  (command "PLAN" "C")
  (princ)
)
(defun C:PW ()
  (command "UCS" "W")
  (command "PLAN" "W")
  (command "ZOOM" ".9X")
  (princ)
)
(defun C:ISO ()
  (command "VPOINT" "1,-1,1")
(princ)
  )
```

These eight macros could be typed into a text editor and saved as a file with a descriptive name such as 3DUTILS.LSP. In fact, that is the name of the file containing these macros that is found on the CD-ROM accompanying this book. Although these macros are intended to speed up work in 3D, their basic structure could easily be applied to other macros.

 Note: Standard AutoLISP files have an LSP extension.

The following exercise demonstrates one method of loading files containing one or more AutoLISP macro definitions.

EXERCISE: LOADING AN AUTOLISP MACRO FILE

1. Start AutoCAD and open Chap07.dwg. (This drawing is contained on this book's accompanying CD-ROM.) The loaded drawing should resemble Figure 7–1.

2. From the Tools menu, select Load Application to display the Load/Unload Applications dialog box as shown in Figure 7–2. (The APPLOAD command also displays this dialog box.)

3. In the Look In input box of the dialog box, find this book's CD-ROM, then find and select the file 3DUTILS.LSP. Select the Load button and note that the 3DUTILS.LSP file is added to the top of the list in the Loaded Applications folder. Also note that an informational message is displayed in the message box near the bottom of the dialog box. Choose Close to exit the dialog box.

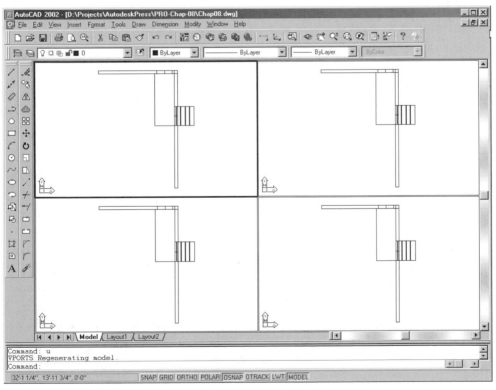

Figure 7–1 *Chap07.dwg.*

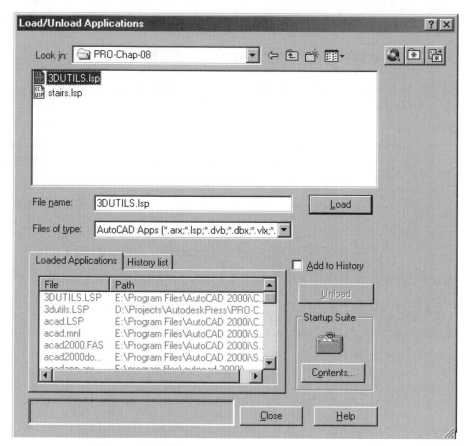

Figure 7–2 *The Load/Unload Applications dialog box.*

4. If necessary, pick in the upper-left viewport to set it current. Test the C:ISO macro by typing **ISO** and pressing Enter. Note that the view in the viewport changes to a standard isometric in accordance with the AutoLISP code in the C:ISO macro.

5. Set the upper-right viewport current. Test the C:RX macro by typing **RX** and pressing Enter. Note that the view changes to a plan view with the UCS rotated positive 90 degrees about the X-axis.

6. Set the lower-right viewport current. First, establish the same view as in the upper-right viewport by running the C:RX macro in this viewport. Then run the C:RY macro by typing **RY** and pressing Enter. Note that the view changes to a plan view with the UCS rotated 90 degrees about the X-axis and 90 degrees about the Y-axis.

7. With the lower-left viewport current, test the C:RZ macro by typing **RZ** and pressing Enter. Note that the view changes to a plan view with the UCS rotated 90 degrees about the Z-axis.

8. Finally, with the lower-left viewport still current, test the C:PW macro by typing **PW** and pressing Enter. Note that the view changes to a World UCS plan view with a 10 percent zoom-out factor.

9. Close AutoCAD *without* saving changes to Chap07.dwg.

Simple AutoLISP keyboard macros such as those contained in 3DUTILS.LSP allow you to perform repetitive command sequences with a minimum number of keystrokes, and they are relatively easy to write. Later in this chapter you will see how more complicated AutoLISP routines and programs take over where macros leave off, offering the ability to receive input from the user and make decisions based on that input.

WORKING WITH THE LOAD/UNLOAD APPLICATIONS DIALOG BOX

In the previous exercise you loaded an AutoLISP file using the Load/Unload Application dialog box. This dialog box provides the graphical interface for AutoCAD's APPLOAD command. The APPLOAD command can be used when you need to load applications that are not automatically loaded when you start AutoCAD. You can use APPLOAD to load applications, unload applications, store a history list of applications you've loaded, and create a startup list of applications to be loaded each time you start AutoCAD. The APPLOAD command was introduced in Release 12, and the graphical interface changed little in appearance and functionality until the appearance of the current Load/Unload Applications dialog box in AutoCAD 2000, when significant functionality and a new visual appearance were added (see Figure 7–2). The main features of this dialog box are outlined as follows:

- **Options at the top of the dialog box**—Derived from the standard file selection dialog box common to many Release 15 file operation dialog boxes in which you can navigate through your directory structure and select files to open. See Chapter 1, Working with OLE Objects, for more details.

- **Load**—Loads or reloads applications that are selected in either the files list or the History List folder. The Load button is unavailable until you select a file that you can load. APPLOAD loads ARX, VBA, LSP, VLX, FAS, and DBX applications.

- **Loaded Applications**—Displays an alphabetical list, by file name, of currently loaded applications. LISP applications are displayed in this list **only if you loaded them in the Load/Unload Applications dialog box.**

- **History List**—Displays an alphabetical list, by file name, of applications that you previously loaded with Add to History selected.

- **Add to History**—Adds any applications that you load to the History List.

- **Unload**—Unloads the selected applications. Unload is not available for all file types. You cannot unload LISP applications, for example.

- **Remove**—Removes the selected applications from the history list. Remove is available only when you select a file on the History List. Remove does not unload an application. The Remove option is also available from a shortcut menu by right-clicking on an application on the History List.

- **Startup Suite**—Contains a list of applications that are started each time you start AutoCAD. Click the Startup Suite icon or Contents to display a secondary dialog box, the Startup Suite dialog box, which displays the contents of the Startup Suite and allows you to add or remove files from the list. You can also add files to the Startup Suite by right-clicking on an application in the History List folder and choosing Add to Startup Suite from the shortcut menu.

Tip: You can drag files into the Loaded Applications files list or into the History List from either the main files list (at the top of the dialog box) or from any application that allows dragging, such as Windows Explorer.

Note: In addition to using the Startup Suite available with the APPLOAD command, you can create a file named ACAD.LSP composed of AutoLISP functions and routines. AutoCAD will automatically load the contents of this file each time AutoCAD is started. The ACAD.LSP file is covered later in this chapter.

CREATING AN AUTOLISP PROGRAM

Now that you have seen how to use AutoLISP in the relatively simple context of a keyboard macro, it is time to progress to a more complex example. AutoLISP *routines* lie a step above macros in complexity—and flexibility. Generally speaking, AutoLISP routines are distinguished by the fact that they will accept input from the user (or from other routines) and are capable of branching in more than one direction based on the input they receive. AutoLISP routines, in other words, exhibit "intelligence"—or at least the ability to make simple decisions. The hallmark of AutoLISP routines, then, is the fact that they almost always request some form of input from the user.

FUNCTIONS, ROUTINES, AND PROGRAMS

There is some confusion regarding the terms *functions, routines,* and *programs.* Generally speaking, functions are operators that perform a task—such as the arithmetic functions of addition or subtraction. AutoLISP has dozens of functions and is one of a few programming languages that allow users to write new functions. Routines are composed of one or more AutoLISP functions—such as the AutoLISP macros (or routines) found earlier in this chapter. AutoLISP functions are the building blocks of all AutoLISP code. If two or more routines are grouped together to accomplish some desirable task, the grouping is commonly referred to as a program. Programs, then, are usually composed of two or more routines. Some routines are rather specialized and are found in specialized programs. Other routines are of a basic, generalized nature and can be used, slightly modified, in a variety of larger programs. This *modular* use of routines provides the AutoLISP programmer with a library of routines that can be utilized over and over again in various program contexts, making the building of larger programs easier. Programming purists may quibble with these definitions and distinctions, but for the purpose of learning AutoLISP, they are functional, workable definitions.

STAIR.LSP—A MODULAR AUTOLISP PROGRAM

The AutoLISP program, STAIR.LSP, is an example of a modular program. STAIR.LSP is included on this book's CD-ROM. A listing of its AutoLISP code follows:

```
;;Get Information
;;Stairs.lsp
;;
;;Get Information
(defun Get-Info ()
  (initget 1 "Standard Modern")
  (setq optn  (getkword "\nStair type: [S]tandard, [M]odern?  "))
  (setq wid   (getdist  "\nEnter step width in inches: "))
  (setq tread (getdist  "\nEnter tread in inches: "))
  (setq rise  (getdist  "\nEnter rise in inches: "))
  (setq num   (getint   "\nNumber of steps: "))
  (setq crnr  (getpoint "\nSpecify lower-left corner of bottom step:
  "))
)
;
;insert proto-step
(defun Proto ()
  (command "zoom" "all")
  (command "vpoint" "-1,-1,1")
  (command "-insert" optn crnr "x" wid tread rise "0")
)
;
;copy steps
(defun Copy-Steps ()
  (repeat (1- num)
    (command "copy" "L" "" (list 0 tread rise) "")
    (command "zoom" "e")
```

```
     )
   )
   ;
   ;make a command
   (defun C:STAIRS  (/ optn tread wid rise num crnr)
     (get-info)
     (proto)
     (copy-steps)
     (princ)
   )
```

 Note: With few exceptions, AutoLISP is not case sensitive. The choice of upper case or lower case is largely a matter of programming style or programmer's choice. Function names are often given an initial capital letter for visual clarity. Likewise, programmers treat symbols and variables differently. Generally, you can disregard case. Punctuation and capitalization in prompts should conform to normal text practice because they are read by the user at the command prompt.

This program consists of three self-contained modular routines:

- Get-Info
- Proto
- Copy-Steps

The purpose of STAIRS.LSP is to draw automatically a set of stairs in 3D. The user supplies the number of steps, as well as the width, tread, and rise of each step. A typical set of steps drawn with STAIRS.LSP is shown in Figure 7–3.

In the sections that follow, the three routines will be briefly discussed so that you can gain a basic understanding of how a number of AutoLISP functions operate, how functions go together to make routines, and how routines fit together to make a simple AutoLISP program. Keep in mind that the intent of this chapter is to provide you with an introduction to AutoLISP not an in-depth tutorial of AutoLISP program building.

Gathering Information with the Get-Info Routine

The first routine in STAIR.LSP, Get-Info, is intended to gather information by requesting input from the user. A series of prompts asks for the style of stair to draw, the tread of each step, the width of the stair, the rise of each step, the number of steps, and the position of the bottom step:

```
   (defun Get-Info ()
     (initget 1 "Standard Modern")
```

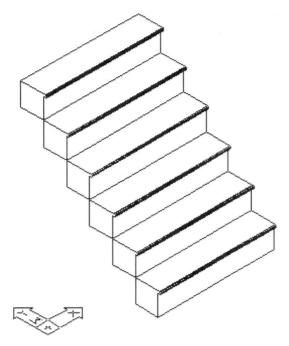

Figure 7–3 *3D stairs drawn with STAIRS.LSP.*

```
(setq optn  (getkword "\nStair type: [S]tandard, [M]odern?  "))
(setq wid   (getdist  "\nEnter step width in inches: "))
(setq tread (getdist  "\nEnter tread in inches: "))
(setq rise  (getdist  "\nEnter rise in inches: "))
(setq num   (getint   "\nNumber of steps: "))
(setq crnr  (getpoint "\nSpecify lower-left corner of bottom step:
            "))
)
```

AutoLISP functions in the *getxxx* family of functions are used to collect this data. Data specifying the dimensions of each step, for example, is collected via AutoLISP's *getdist* function. The *getdist* function accepts either typed distance input or distance input derived from screen picks. As with all of the *getxxx* functions, you can include a prompt as a quoted text string. The first *getdist* function in the Get-Info routine prompts you for the step width. The distance supplied in response to this prompt is stored as a *variable*—an arbitrary symbol—by the *setq* function. Although there are few rules governing the choice of symbols, using a symbol that somehow represents the data that the variable stores is helpful in following the sense of the routine. In this instance, the programmer chose the variable *wid* because the variable holds the value of the width of the stair.

In a similar manner, *getdist* functions, with appropriate prompts, are used to obtain the tread and rise distances. Again, a *setq* function stores the input distances of each parameter in the variables *tread* and *rise*, respectively.

A different *getxxx* function is used to obtain the number of stairs to be drawn. Because the number of stairs is represented not by a distance but rather by a whole number, the *getint* (for "get integer") function is used. Again, an appropriate prompt is included, and the input is stored in the variable *num* (for number) by the *setq* function.

Points are sets of three numbers representing the X-, Y-, and Z-coordinates of a point. A *getpoint* function is therefore used next to obtain the lower-left starting point of the stairway. Points are usually supplied by picking a point on the screen, although data can be typed in at the command line. The point is stored in the variable *crnr* (for corner).

You may often want to restrict the input to a particular *getxxx* function. For example, you may want to disallow a "zero," a negative number, or even an Enter. Or you may wish to restrict input to one of several permissible choices (usually strings). The *initget* function provides this flexibility. For example, an *initget* function is used in the Get-Info routine to obtain the name of the type of step. Because a block will be used by STAIRS.LSP to draw each individual step, it is important that the user knows the choices available and does not enter an erroneous choice. The programmer made two styles of steps available: Standard and Modern. The *initget* function, therefore, restricts the user's response to the very next *getxxx* function by setting up a list of permissible choices—Standard and Modern. Because the choices are specified with initial capital letters, users may merely designate their choice by typing the capitalized letter instead of the entire name. This conforms to the convention used in AutoCAD prompts and is therefore understood by users.

The *initget* function acts somewhat as a filter. Its restrictions apply only to the next *getxxx* function, which in the case of Get-Info is a *getkword* function. The term "kword" in this context stands for keyword, and the keywords allowed in response to a *getkword* function are those initialized by the preceding *initget*. Get-Info offers two keywords: Standard and Modern. The user selects one of these in response to the *getkword* prompt, and the choice is stored by the *setq* function in the variable *optn* (for "option"). When the Get-Info routine is completed, six pieces of data are stored in six variables. These variables will be used by the other two routines in the program.

Putting the Variables to Work with the Proto Routine

The Proto routine is little more than an AutoLISP triple macro. The big difference between Proto and a typical macro is that Proto uses data collected outside the routine itself. Proto is composed of three functions. The first function executes AutoCAD's ZOOM command and specifies the All option. In a plan view, a ZOOM/All will zoom the drawing to either the drawing's limits or its extents, whichever is greater. In a 3D view, a ZOOM/All is equivalent to a ZOOM/Extents. This zoom is included to provide adequate "room" in the drawing for the block insert. The second function in Proto is a specific change to a non-plan view. The VPOINT command is called with a SW Isometric viewpoint vector:

```
(defun Proto ()
  (command "zoom" "all")
  (command "vpoint" "-1,-1,1")
  (command "-insert" optn crnr "x" wid tread rise "0")
)
```

The last function in the Proto routine calls the command line version of the INSERT command. The command line version is required because AutoLISP cannot deal directly with dialog boxes. In AutoCAD 2000/2002, to invoke the command line version of the INSERT command, the command's name must be preceded by a dash.

 Note: Prior to AutoCAD 2000, the command line version of the INSERT command was either the only version or the default version and was called with the name "insert." In Releases 12, 13, and 14, the dialog box version was called with the name "ddinsert." Beginning with AutoCAD 2000, the dialog box version is the default version called with "insert," and the command line version is made available with "-insert."

A review of the prompt sequence for the command line version of INSERT command will make the third function's call to "-insert" clear. The sequence of prompts with the responses used by the Proto routine follows.

Enter block name or [?]:

The variable *optn* is supplied here. It is not enclosed inside double quotation marks because the *getkword* function stores it as a string automatically. Specifically, variable *optn* will contain either the string "standard" or "modern." Therefore, the drawing must also contain blocks by this name.

Specify insertion point or [Scale/X/Y/Z/Rotate/Pscale/...

The Get-Info variable *crnr* is supplied in answer to this prompt.

Enter X scale Factor, specify opposite corner, or

You want to be able to specify different scale factors for X, Y, and Z during insertion of the "Proto" step block, so an X enclosed in double quotation marks (as a string) is supplied in response to this prompt.

Specify X scale factor or [Corner]...

The X scale factor corresponds to the stair width, which is held in the *wid* variable.

> Specify Y scale factor...

The Y scale factor corresponds to the stair tread held in the *tread* variable.

> Specify Z scale factor...

The Z scale factor is the rise of the Proto step and is held in the *rise* variable.

> Specify rotation angle...

The last prompt asks for rotation angle. This will always be zero degrees. The value "0" is specifically quoted.

This completes the Proto routine. If the program were to halt at this point, you would see either the Standard or Modern block inserted with the lower-left corner at the point specified in the Get-Info routine. The block would be differentially scaled so that the width of the step would equal the value held by variable *wid*, the depth (or Y distance) of the step would equal the value held in variable *tread*, and the height of the step would equal the value held by variable *rise*.

 Note: In order to take advantage of the INSERT command's ability to scale differentially in the X, Y, and Z directions, the blocks used for STAIRS.LSP must be defined with unit height, width, and depth.

Building the Stairs with the Copy-Steps Routine

The Copy-Steps routine is the heart of the STAIRS.LSP program. Once the Proto routine has inserted the first step at the proper X, Y, and Z scale factor, the Copy-Steps routine takes over to build the stairway:

```
(defun Copy-Steps ()
  (repeat (1- num)
    (command "copy" "L" "" (list 0 tread rise) "")
    (command "zoom" "e")
  )
)
```

Copy-Steps contains two functions, each enclosed by a *repeat* function loop. The number of "repeats" is equal to the number of stairs stored in the *mum* variable minus 1—because the first step is already in place.

The first function enclosed in the repeat loop consists of a macro-like execution of the COPY command:

```
(command "copy" "L" "" (list 0 tread rise) "")
```

Again, a review of the COPY command's prompt sequence shows how Copy-Steps builds the stairway. Here is the sequence:

```
Select objects:
```

As with most AutoCAD edit commands, you can specify the last object drawn at the "Select objects" prompt. A quoted "L" is first supplied in answer to the prompt. This refers to the block inserted in the Proto routine.

```
Select objects:
```

Close the repeating "Select objects" prompt by supplying the macro equivalent of a keyboard Enter—a pair of double quotation marks.

```
Specify base point or displacement, or [Multiple]:
```

We need to displace the copy by an amount equal to the stair tread and rise, that is the values held in variables *tread* and *rise*. No displacement in the X direction is wanted. We can supply this displacement by listing an X value of zero, a Y value of *tread*, and a Z value of *rise*. In AutoLISP, the most direct method of supplying this displacement is to use the AutoLISP *list* function:

```
(list 0 tread rise)
```

This response is followed by a pair of double quotation marks (to indicate an Enter) to complete the displacement specification and end the COPY command.

The second function contained in the repeat loop is a straightforward execution of the ZOOM command with an Extents parameter. Performing this zoom after each call to the COPY command will ensure that all the copied steps are displayed in the current viewport. This is necessary because the Last option in the "Select objects" prompt of the COPY command requires that the object be visible.

The Copy/Last and Zoom/Extents events are repeated (variable *num* - 1) times to yield the correct number of steps.

Lastly, we group the three STAIRS.LSP routines together:

```
(defun C:STAIRS (/ optn tread wid rise num crnr)
   (get-info)
   (proto)
   (copy-steps)
   (princ)
)
```

As with AutoLISP macros, a "C:" prefix to a user-defined function such as STAIRS.LSP elevates the function or program to the level of an AutoCAD command that can be entered directly at any command prompt. Immediately following the naming of the program is a list of all the variables used in the constituent routines. Listing the variable in this manner effectively deletes them from AutoLISP's memory, freeing the memory space for other variables. The last function in the C:STAIRS function is a call to another AutoLISP function, *princ*. The sole purpose of including this function is to eliminate the "nil" that would otherwise be printed to the command line upon completion of C:STAIRS.LSP. This "nil" is visually unwanted and intimidating to many users.

In the following exercise you will load and run STAIRS.LSP in this chapter's Chap08.dwg.

 Note: STAIRS.LSP attempts to insert a block. The block must be defined in the drawing or the Proto routine will fail. The provided file Chap08.dwg contains the blocks required for STAIRS.LSP.

EXERCISE: LOADING AND RUNNING THE STAIRS.LSP PROGRAM

1. Open Chap08.dwg. Type **AP** and press Enter to start the APPLOAD command and display the Load/Unload Applications dialog box as shown in Figure 7–4.

2. In the Look In portion of the dialog box, find this book's CD-ROM, and then find and select the file STAIRS.LSP. Select the Load button and note that the STAIRS.LSP file is added to the top of the list in the Loaded Applications folder. Also note that an informational message is displayed in the message box near the bottom of the dialog box. Choose Close to exit the dialog box.

3. If necessary, click in the upper-left viewport to set it current. To configure a single viewport, from the View menu, choose Viewports>1 Viewport. Then, from the View menu, select Named Views. In the View dialog box, select view AAA and Set Current. Select OK to close the dialog box and restore view AAA. Your drawing should resemble Figure 7–5.

    In the following steps you will test the STAIRS.LSP program that you loaded in step 2.

4. Type **STAIRS** and press Enter. Type responses to the STAIRS.LSP prompts as follows:

    Stair type: [S]tandard, [M]odern? **S**

    Enter step width in inches: **36**

    Enter tread in inches: **8**

    Enter rise in inches: **6**

    Number of steps: **12**

    Specify lower-left corner of bottom step: **17', 7'**

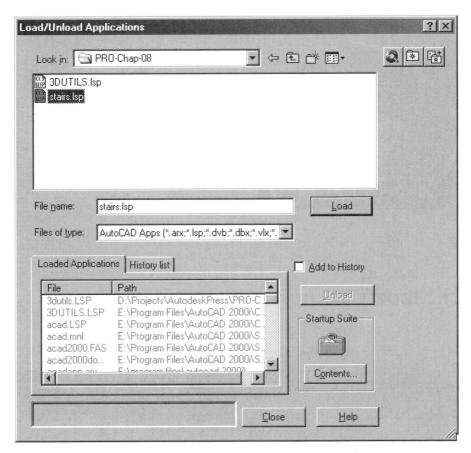

Figure 7–4 *The Load/Unload Applications dialog box.*

5. Use the HIDE command (View>Hide) to obtain a hidden line view (see Figure 7–6).

6. Close Ch07.dwg *without* saving changes.

STAIRS.LSP is useful here as an example of the kinds of sophisticated programming you can perform in AutoLISP. Several improvements, such as the ability to control the direction of the stairway, could be added to STAIRS.LSP to make it even more useful in a typical 3D modeling or drafting environment. However, even though STAIRS.LSP is not a complex program, its ability to accept user input places it ahead of the simpler, but useful, AutoLISP macros and shows why and how AutoLISP is such a popular means of customizing and extending AutoCAD.

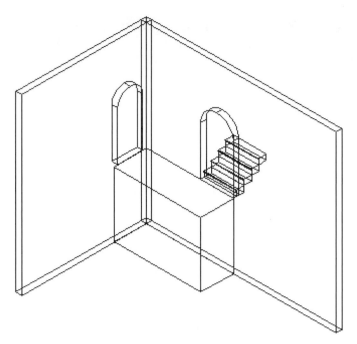

Figure 7–5 *Ch07.dwg with view AAA restored.*

UNDERSTANDING ACAD.LSP AND ACADDOC.LSP

If you want to load a group of AutoLISP routines every time you start AutoCAD, you can place the routines in a standard (ASCII) text file and save the file with the name ACAD.LSP. When AutoCAD starts, it searches its library path for a file named ACAD.LSP. If it finds such a file, its contents are loaded into memory. By default, ACAD.LSP is loaded only once: when AutoCAD starts.

If you want to load a group of AutoLISP routines every time you start a new drawing (or open an existing drawing), you should create a file named ACADDOC.LSP and place it in AutoCAD's library path. Each time AutoCAD opens a drawing, it searches the library path for a file named ACADDOC.LSP and, upon finding one, loads the contents of the file into memory.

You can have ACAD.LSP load with every drawing by selecting the option Load ACAD.LSP with Every Drawing in the System folder of the Options dialog box. If this option is not checked, only the ACADDOC.LSP file is loaded into all drawing files. Clear this option if you do not want to run certain LISP routines in specific drawing files. You can also control the option Load ACAD.LSP with Every Drawing by using the ACADLSPASDOC system variable.

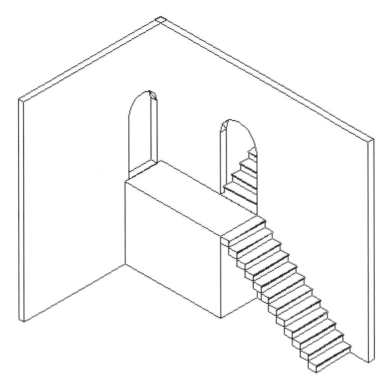

Figure 7–6 *Stairs drawn with STAIRS.LSP.*

By controlling the location of the ACAD.LSP and ACADDOC.LSP files in the AutoCAD library path, you can control whether or not these files are found (and loaded). AutoCAD will load the first ACAD.LSP and ACADDOC.LSP it encounters, so by utilizing AutoCAD Profiles and project directories, you can control which among several ACAD.LSP files, for example, is found first and loaded.

SUMMARY

AutoLISP is one of the oldest and most popular programming interfaces in AutoCAD. Although it is a "full-blown" modern programming language, AutoLISP is not difficult to learn, and even a basic understanding of its usage within AutoCAD can give you the ability to write simple programs and macros that can save significant amounts of time by allowing you to customize AutoCAD according to the way you work.

CHAPTER   8

Customizing with DIESEL

DIESEL is an acronym for Direct Interpretively Evaluated String Expression Language. Like AutoLISP and Visual LISP, DIESEL is an AutoCAD Application Programming Interface, or API. Unlike AutoLISP and Visual LISP, which are derive from and dialects of the larger, well-known LISP language, DIESEL is unique to AutoCAD. Although DIESEL resembles LISP in several important ways, it is not dependent upon AutoLISP for its operation. On the one hand, a knowledge of AutoLISP will make learning and using DIESEL easier. Similarly, having a working knowledge of DIESEL will help anyone wanting to learn AutoLISP or Visual LISP. This chapter will introduce you to DIESEL and some of its capabilities.

This chapter will cover the following topics:

- Understanding the mechanics of DIESEL
- Learning DIESEL functions
- Writing DIESEL expressions
- Debugging DIESEL
- Using the MODEMACRO system variable
- Using DIESEL with AutoLISP

INTRODUCING DIESEL

Although DIESEL is not as fast, robust, or flexible a language as AutoLISP is, it does have several distinct capabilities not available with AutoLISP. First of all, you can configure the MODEMACRO system variable using a DIESEL expression. The content of the MODEMACRO system variable is displayed in a left-aligned pane in the status bar at the bottom of the AutoCAD window, as shown in Figure 8–1. Although the MODEMACRO system variable, by default, has no initial value, it can be made to display useful information using DIESEL. This is historically the primary use of DIESEL and it is the application discussed in this chapter.

Although it is beyond the scope of this chapter, DIESEL can also be utilized in pull-down menu labels. A DIESEL expression incorporated in this manner is read dynamically each time the pull-down menu is activated, allowing label items to change their appearance (e.g., to be displayed as checked or unchecked) according to the conditions tested for by the underlying DIESEL expression.

Lastly, DIESEL expressions can also be placed in the body of menu macros. When such a macro is executed, the DIESEL expression itself is evaluated. This allows you to add a degree of intelligence to menu macros without resorting to the use of AutoLISP.

 Note: As discussed in Chapter 7, Introduction to AutoLISP Programming, Visual LISP is an extension of AutoLISP. In this chapter the two names are used synonymously.

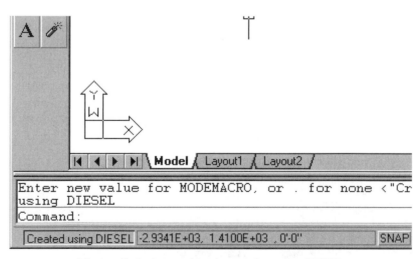

Figure 8–1 *A customized status line using DIESEL.*

Note: Like AutoLISP, DIESEL code is enclosed in pairs of parentheses with nesting not only allowed but frequently utilized. Any written DIESEL code that is intended to be evaluated and whose output will be employed for some useful purpose is referred to as a DIESEL expression.

THE MECHANICS OF DIESEL

Before beginning to work with DIESEL you need to understand some of the basic characteristics of this specialized AutoCAD programming language. If you are familiar with AutoLISP, you will find DIESEL similar in most respects—especially in DIESEL's use of parentheses and its basic syntax. If, on the other hand, you have little or no experience with AutoLISP, learning the basics of DIESEL will serve as a good foundation for learning AutoLISP. Neither of the two languages presupposes knowledge of the other.

 Note: For information about AutoLISP, refer to Chapter 7, Introduction to AutoLISP Programming.

DIESEL is a macro string evaluation language. It consists of a relatively small group of some twenty-eight functions that manipulate and process data in *string* form. Strings are a data type common to most programming languages. Generally, strings are anything enclosed within a pair of double quotation marks. DIESEL expressions generate string output.

As mentioned before, both AutoLISP and DIESEL expressions are delimited by parentheses, and the first element following the open, or left, parenthesis is the name of the function. Generally, the function names in DIESEL are the same as their counterparts in AutoLISP. As in AutoLISP, DIESEL function names are not case sensitive, and spaces that are not between commas are not read. All DIESEL expressions have the following general form:

```
$(function,argument1,argument2,…)
```

The following features are distinctive to DIESEL:

- A dollar sign ($) immediately precedes each opening (left) parenthesis.

- Each element (function or argument) within a pair of parentheses is considered a member of a list and is followed by a comma.

- Two commas appearing consecutively with nothing between them is considered a null, or empty, string.

- Because a comma delimits each element in a DIESEL expression, spaces that appear *between* two consecutive commas are interpreted literally and appear in the result.

Here is a typical DIESEL function call:

```
$(getvar,snapunit)
```

In this example, the DIESEL function *getvar* returns the value of the AutoCAD system variable SNAPUNIT. The string returned contains the X and Y values of the current snap increment.

 Note: Unlike AutoLISP, you cannot enter DIESEL expressions, such as the preceding one, directly at the command prompt. Later in this chapter an AutoLISP function is given that allows you to enter DIESEL expressions at a command prompt.

Here is another DIESEL expression using the DIESEL function for addition:

```
$(+,2,5)
```

The sum of the integers 2 and 5 would be returned as a string.

The following expression demonstrates the nesting of DIESEL expressions:

```
$(rtos,$(getvar,ltscale),2,4)
```

In this expression, the current value of the LTSCALE system variable is first evaluated using the *getvar* function. This value, a real number, is then used as an argument for the *rtos* (real-to-string) function, formatted in decimal units to four places.

LEARNING DIESEL FUNCTIONS

It is beyond the intent of this chapter to present all the DIESEL functions with their required arguments and examples of their implementation. AutoCAD, however, contains a complete catalog of the DIESEL functions in its Help facility, as shown in the following example.

ACCESSING DIESEL FUNCTION DEFINITIONS

1. In AutoCAD, with no command in progress, press F1.

2. In the Contents folder, expand the Customization Guide topic, as shown in Figure 8–2.

3. From the list of topics, expand the DIESEL—String Expression Language topic.

4. Expand the Catalog of DIESEL String Functions.

5. Browse through the alphabetical list of DIESEL functions as shown in Figure 8–3.

PRACTICING DIESEL EXPRESSIONS

As mentioned earlier, unlike AutoLISP expressions, DIESEL expressions cannot be entered directly on the command line. There is a "workaround" for this limitation, however. The AutoLISP function *menucmd* allows the direct input and interpreta-

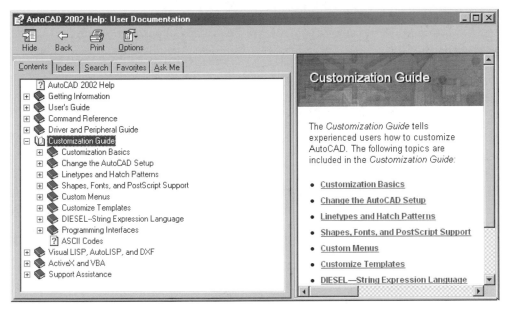

Figure 8–2 *Expanded Customization Guide topic.*

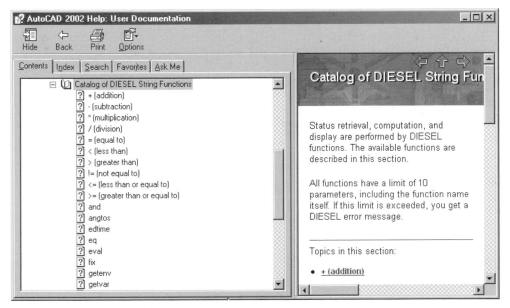

Figure 8–3 *AutoCAD Help's catalog of DIESEL functions.*

tion of DIESEL expressions. The following AutoLISP user-defined function adds the DIESEL command to AutoCAD. Once loaded, this command lets you enter

DIESEL expressions on the command line. The expressions are evaluated and the result is displayed. This allows you to easily test and "experiment" with DIESEL expressions.

```
(Defun C:DIESEL (/ dsl)
  (while (/= "" (setq dsl (getstring T "\nDiesel>: ")))
    (print (menucmd (strcat "M=" dsl)))
  )
(princ)
)
```

Note: This *C:DIESEL* AutoLISP function appears on this book's CD as the file DIESEL.LSP. You may want to copy this file from the CD to a directory in your current AutoCAD library path, such as the AutoCAD\Support directory under your AutoCAD installation.

In the following exercise you will load the *C:DIESEL* function and then use it to test several DIESEL functions.

EXERCISE: TESTING DIESEL FUNCTIONS WITH *C:DIESEL*

1. From the Tools menu, select AutoLISP>Load to display the Load/Unload Applications dialog box as shown in Figure 8–4.

2. In the upper portion of the dialog box, navigate to the file DIESEL.LSP on this book's CD. (If you have copied this file to your AutoCAD directory structure, navigate there.)

3. With the DIESEL.LSP file appearing in the File Name input window, select the Load button. The statement "Diesel.lsp successfully loaded." is displayed in the message box (see Figure 8–5). Choose Close to close the dialog box.

4. At the command prompt, type **DIESEL** and press Enter. AutoCAD displays the DIESEL>: prompt.

5. Enter **$(getvar, dwgname)**. DIESEL returns the following:

   "Drawing1.dwg"

   (If your current drawing name is different, that name will appear.)

6. Press Enter. The standard AutoCAD command prompt returns.

7. Type **Circle** and press Enter. Draw a circle of any size.

8. At the command prompt again type **DIESEL** and press Enter. The DIESEL>: prompt returns.

9. Enter **$(getvar, circlerad)**. DIESEL returns the following:

   "2.95"   (or whatever the radius of your circle is)

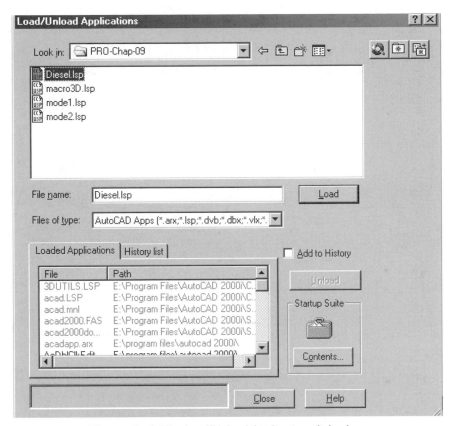

Figure 8–4 *The Load/Unload Applications dialog box.*

10.   Enter **$(getvar lastpoint)**. DIESEL returns the following:

"2.4, 4.5, 0" (or whatever the center of your circle is)

11.   Press Enter. The standard AutoCAD command prompt returns.

Once it is loaded into memory, you can call the *C:DIESEL* AutoLISP function by typing **DIESEL** at the command prompt. Pressing Enter at any DIESEL>: prompt causes the standard AutoCAD command prompt to return.

The *C:DIESEL* function is a convenient and helpful aid to use while learning and testing DIESEL expressions. The output of *C:DIESEL* will, in most cases, also report typing or syntax errors as well as valid output. This capability is a helpful learning tool. For example, the following *C:DIESEL* input generates a corresponding error output:

```
DIESEL>: $(+ 3,5,2)
" $(+ 3)?? "
```

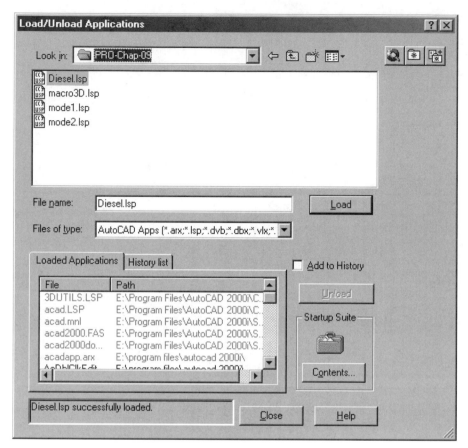

Figure 8–5 *Loading the DIESEL.LSP file.*

The error message points to the problem: The comma following the function (+, in this example) is absent and the error message states that the function $(+ 3) is unknown. Correcting the expression to the following results in a valid output:

```
DIESEL>: $(+,3,5,2)
"10"
```

The *C:DIESEL* function also demonstrates that AutoLISP and DIESEL can co-exist. Keep in my, however, that *C:DIESEL* is merely a handy learning and development aid; it performs no otherwise useful work.

DEBUGGING DIESEL EXPRESSIONS

As with AutoLISP expressions, DIESEL expressions can quickly become quite complex, with expressions nested within other expressions. Typing and syntactical errors become common and difficult to locate visually. A missing comma, for example, can bring expression evaluation quickly to a halt. Fortunately, there is a built-in method of tracing DIESEL expressions step by step, helping to isolate such errors. AutoCAD's largely undocumented MACROTRACE system variable allows you to trace even the most complex DIESEL expression, pinpointing any errors. When you activate the MACROTRACE system variable (i.e., set it to a value of 1), the evaluation of all DIESEL expressions is *traced*, with each DIESEL expression (including any nested expressions) displayed as it is evaluated. The following exercise demonstrates how the MACROTRACE system variable works.

 Note: The following example assumes that you have previously loaded the AutoLISP function *C:DIESEL*. This function is contained on the book's CD as the file DIESEL.LSP. Refer to the preceding example for loading instructions. Once *C:DIESEL* is loaded you can proceed with the following example.

EXERCISE: TRACING DIESEL EXPRESSIONS

1. At an empty command prompt, enter **MACROTRACE**. Answer the following prompt as shown:

   Enter new value for MACROTRACE <0>: 1

   Then press Enter.

2. At the command prompt, type **DIESEL** and press Enter. The following prompt appears:

   DIESEL>:

3. Carefully enter the following DIESEL expression:

   ```
   $(if,$(=,$(+, 4,3),$(-, 10,3)),The Answer is Ten!)
   ```

4. Press Enter.

   The following expression trace appears:

   ```
   Eval: $(IF, $(=,$(+, 4,3),$(-, 10,3)), The Answer is Ten!)
   Eval: $(=, $(+, 4,3), $(-, 10,3))
   Eval: $(+,  4, 3)
   ===>  7
   Eval: $(-,  10, 3)
   ===>  7
   ===>  1
   ===>  The Answer is Ten!
   ```

5. Enter the following DIESEL expression. (The syntax error is intentional.)

   ```
   $(*,10,$(+ 2,3))
   ```

The following expression trace appears:

```
Eval: $(*, 10, $(+ 2,3))
Eval: $(+ 2, 3)
Err:    $(+ 2)??
Err:    $(*,??)
```

6. Enter the following DIESEL expression:

    ```
    $(=,3,3)
    ```

    The following expression trace appears:

    ```
    Eval: $(=,  3, 3)
    ===>  1
    "1"
    ```

7. Enter the following DIESEL expression:

    ```
    $(=,3,5)
    ```

    The following expression trace appears:

    ```
    Eval: $(=,  3, 5)
    ===>  0
    "0"
    ```

8. At the DIESEL>: prompt press Enter to return to a standard AutoCAD command prompt.

9. Type **MACROTRACE** and press Enter. Answer the following prompt as shown:

    Enter new value for MACROTRACE <1>: **0**

    Then press Enter.

In step 3 of the preceding example you entered a complex DIESEL expression, nested two levels deep. This expression can be translated into non-code language as follows: "If the sum of the integers 4 and 3 is equal to the difference of the integers 10 and 3, return the string 'The Answer is Ten!'" By following each successive line of the trace, you can see MACROTRACE's evaluation of each of the nested expressions. The addition expression is first evaluated and its value is "printed." Next, the subtraction expression is evaluated and its value is printed. The entire IF expression is next evaluated and its value, 1, is printed. Finally, the "then" clause of the IF expression is printed, or returned. In this case the string "The Answer is Ten!" is returned because the value of the IF expression is non-nil, or True.

It is important to note that in DIESEL, expressions that are evaluated as True (or non-nil) return the string "1," while nil or False expressions return the string "0." This is an important distinction from AutoLISP (and most other programming languages), which returns the value T (for True) and nil for False when it evaluates True/False expressions. This behavior is also demonstrated in steps 6 and 7 of the preceding example. This difference will become advantageous when DIESEL is used to evaluate various system variables as test predicates, because system variables often are either "on" or "off"— or "1" or "0," respectively.

In step 5 of the preceding example, the entered expression contains a syntactical error in that the $(+ function is not immediately followed by a comma. This is shown in the trace, which reports an error when $(+ 2 is encountered.

There are four error messages generated by MACROTRACE to help pinpoint the source or nature of the error. These messages are listed in Table 8–1.

The MACROTRACE system variable is undocumented; you will not find it listed in AutoCAD's Help facility. But it is available in Release 13, Release 14, and AutoCAD 2000/2002. In combination with the *C:DIESEL* AutoLISP function, it is a valuable learning and error tracing tool.

Table 8–1: *MACROTRACE Error Messages*

Error Message	Description
$?	Syntax error
$?(func,??)	Incorrect argument to a function
$(++)	Output string too long
$(func)??	Function unknown

CREATING CUSTOM STATUS LINES WITH DIESEL

DIESEL was first introduced in AutoCAD Release 12. Its primary purpose was to provide a means of customizing AutoCAD's mode status line. The mode status line in AutoCAD is the bar that appears at the bottom of the AutoCAD application window. Without any customization, the status line appears as shown in Figure 8–6.

The mode status line is a valuable resource that allows the user rapid access to frequently used functions, such as GRID and SNAP settings. The status line also contains the coordinate readout panel that can be set to give a continuous readout of the screen cursor's current position or relative coordinates from a picked point. An additional, user-defined panel can be added immediately to the left of the coordinate readout panel. This customized—or expanded—mode status line can be configured to display a wide variety of useful information using DIESEL expressions or a combination of AutoLISP and DIESEL expressions. The content of any customized status line is controlled by the MODEMACRO system variable.

THE MODEMACRO SYSTEM VARIABLE

The MODEMACRO system variable displays a string of text, usually written in DIESEL or a combination of DIESEL and AutoLISP language, on the status line. Unlike most other AutoCAD system variables that report, or "hold," data, such as the current drawing name or the on/off status of the ortho mode, MODEMACRO can be made to contain virtually any string of alphanumeric characters you devise.

At startup, the value of the MODEMACRO system variable is set to the so-called "null," or empty, string. In other words, it contains nothing. In this condition the coordinate readout panel appears at the extreme left of AutoCAD's status line as shown in Figure 8–6. To create a customized panel, you give the MODEMACRO system variable a string value. It can be set to any string (i.e., text) value whose length is limited only by the size of the AutoCAD window. In practice, a very long string assigned to MODEMACRO would force the standard AutoCAD panels completely off the display.

To set the MODEMACRO system variable, you enter **MODEMACRO** and press Enter at the command prompt, or use the SETVAR command in conjunction with an AutoLISP expression. In the following exercise you will see how the MODEMACRO system variable works by entering a static status-line text display.

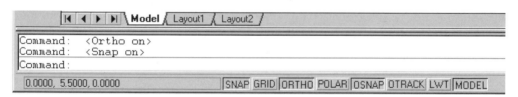

Figure 8–6 *AutoCAD's default mode status line.*

EXERCISE: ASSIGNING A STATIC VALUE TO MODEMACRO

1. At the command prompt, enter MODEMACRO. Answer the following prompt as shown:

   Enter new value for MODEMACRO, or . for none <"">: **Dave's Macro Boutique.**

   Then press Enter (see Figure 8–7).

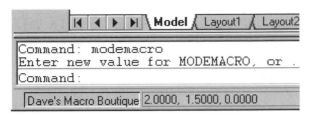

Figure 8–7 *Displaying a static mode status line.*

2.   Press Enter to display the MODEMACRO prompt again. Answer the prompt by typing a period (.) and pressing Enter. Note that the macro status line has been removed.

As you can see, assigning a value to MODEMACRO is not difficult. If you type a string at the MODEMACRO prompt, that string remains displayed on the status line for the remainder of your AutoCAD session or until you change it. It is not saved anywhere, so you must type the message every time AutoCAD is started. The message "Dave's Macro Boutique" is not very informative; it is static and conveys little with respect to the opened, active drawing. Fortunately, there are ways to have MODEMACRO interactively display more useful information. Any of the many AutoCAD system variables, for example, can be displayed using DIESEL expressions. In the following exercise you will use DIESEL to display the value of the current drawing's user elapsed timer.

EXERCISE: ASSIGNING A DYNAMIC VALUE TO MODEMACRO

1.   At the command prompt, enter MODEMACRO. Carefully answer the following prompt as shown:

Enter new value for MODEMACRO, or . for none <"">: **Elapsed time = $(rtos, $(*, 24, $(getvar, tdusrtimer)),2,2) Hours**

Then press Enter.

2.   AutoCAD displays the value of the current drawing's timer on the mode status line (see Figure 8–8).

Note: For more information about AutoCAD's timer functions, see Help under the TIME command and the system variable TDUSRTIMER.

This mode status line is more meaningful. First of all it conveys information that may be of interest to an AutoCAD user—namely, the reading of AutoCAD's internal, user-controlled, elapsed-time timer. This value is stored in the TDUSRTIMER system variable and is therefore available to the DIESEL function *getvar*. Secondly,

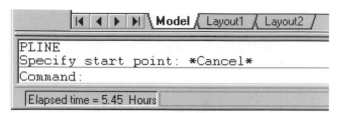

Figure 8–8 *Displaying a dynamic mode status line.*

the display is dynamic; that is, it gets updated whenever user input is supplied in the form of the Enter key or the Esc key being pressed.

The DIESEL expression used in the last example is nested three levels deep and also contains pure text elements. The expression starts with the pure text, "Elapsed time =." The nested DIESEL expression begins with the function *rtos*, which prints a real number in a specified format and precision. The *rtos* function has the following general format:

```
$(rtos, value [,mode, precision])
```

The *value* argument used in the example is itself a nested expression that multiplies the value contained in the system variable TDUSRTIMER by 24 and then expresses the result in decimal mode (mode 2) to a precision of two decimal places. The TDUSRTIMER variable holds the elapsed timer data in days or fractions thereof.

Lastly, the pure text string "Hours" is added at the end.

The MODEMACRO string entered in the preceding example is typical of the information you can format to appear on the mode status line. Virtually all data known to AutoCAD through its system variables can be displayed. In addition, DIESEL offers a function, *edtime*, that greatly facilitates the display of the current time maintained by your computer.

USING AUTOLISP WITH DIESEL

If you typed the DIESEL expressions found in the preceding example, you discovered that entering DIESEL and text manually at the MODEMACRO prompt is a tedious and error-prone process. In addition, you have to manually enter any customized status-line macro display every time you start AutoCAD. Using AutoLISP to format, record, and load your MODEMACRO strings makes this procedure easier, faster, and error-free.

You can assign a value to the MODEMACRO system variable using the AutoLISP function *setvar*. The following user-defined AutoLISP function takes the DIESEL/

text input in step 1 of the preceding example and automatically sets the value of MODEMACRO:

```
(defun C:mode1 ()
 (setvar "MODEMACRO"
    (strcat
 "Elapsed timer: $(rtos,
$(*,24,$(getvar,TDUSRTIMER)),2,2) Hours "
    )
 )
)
```

 Note: The MODEMACRO input code for this function is visually broken down into two relatively short strings for clarity. See Chapter 7, Introduction to AutoLISP Programming, for more about AutoLISP. The code for this macro is contained on the book's CD as the file MODE1.LSP.

Type this code into any text editor, such as Windows Notepad, and save it as MODE1.LSP in your AutoCAD installation's AutoCAD\Support directory. In the following exercise you will load and run this function.

EXERCISE: LOADING AND RUNNING AN AUTOLISP MODEMACRO

1. At the command prompt, enter the text shown in bold and press Enter as indicated:

   Command: **(load "mode1")** and press Enter

   C:MODE1

   Command: **mode1** and press Enter

2. Observe the mode status line. It should appear as shown in Figure 8–8. (The actual time shown will vary depending on the drawing you have open and whether the elapsed time timer is activated.)

3. To return to the standard AutoCAD mode status line, enter **MODEMACRO** at the command prompt, answer the prompt by typing a period (.), and then press Enter.

Alternatively, you can place the AutoLISP code for the MODE1.LSP AutoLISP function in a file named ACAD.LSP. This will cause the *MODE1* function to be evaluated and loaded into memory every time AutoCAD is started. Then, to display the *MODE1* status line, you will only need to type **MODE1** and press Enter at the command prompt. (See Chapter 7, Introduction to AutoLISP Programming, for more information about the ACAD.LSP file.)

SUMMARY

In this chapter you used DIESEL, the string expression language embedded inside AutoCAD, to compose a customized mode status line, providing a more informative, dynamic user interface. The chapter's examples provided you with the basic principals of how DIESEL can be used to extract useful information and display it in a convenient manner. Using these examples, you can build customized status line displays that can make your particular AutoCAD work more efficient.

CHAPTER 9

Introduction to Visual LISP

INTRODUCTION AND BRIEF HISTORY

AutoLISP is a subset of the popular LISP programming language developed in the late fifties. LISP, which is an acronym for LISt Processing, is one of the few programming languages from that era still in active use today. It was developed principally for use in Artificial Intelligence (AI) applications and is still popular in the AI community. It was chosen as the initial AutoCAD programming language for two important reasons: LISP structurally relies upon lists to contain its data types, and most of CAD relies upon lists of coordinates or data that can be easily expressed in list form. Just as important, LISP, and by extension, AutoLISP, are uniquely suited for the relatively unstructured design environment of CAD projects, which typically involve repeatedly trying different solution to the design process.

Beginning with AutoCAD Release 14.01, the AutoLISP programming language embedded in AutoCAD was greatly expanded after having remained essentially unchanged since its introduction in Release 2 in 1986. In the years since 1986, AutoLISP has become widely popular and accepted as an easy-to-learn, easy-to-use customization language for use in AutoCAD. Over 1.5 million AutoCAD customers currently use AutoLISP either directly or through third-party applications integrated with AutoCAD. Despite its popularity and adoption by the AutoCAD user and developer communities, however, AutoLISP began to show some serious limitations as AutoCAD and the machines on which it runs improved in functionality and speed.

Beginning with AutoCAD Release 14, a new implementation of AutoLISP, termed Visual LISP, was introduced. VISUAL Lisp can be best described as a complete development environment for creating applications and customizing AutoCAD using the LISP programming language. Visual LISP is also an extension and enhancement to the "old" AutoLISP found in Release 14 and earlier releases. Visual LISP provides greater functionality, improved productivity and performance, increased security, and a new Integrated Development Environment (IDE). It is important to realize that if you are already familiar with AutoLISP, the transition to Visual LISP will be virtually seamless. Although there are new functions in Visual LISP, the "old" AutoLISP has been essentially retained, and your investment in learning AutoLISP can be easily transported into Visual LISP.

Much of the increased functionality and the improved productivity of Visual LISP are centered upon the fact that it is itself an Object ARX application and it has a completely new LISP interpreter. It greatly expands AutoCAD's ActiveX support by including additional ActiveX objects and events. While many of the advanced aspects of Visual LISP are outside of the scope of this chapter, Visual LISP also includes several new elements that aid in the writing and debugging of AutoLISP/Visual LISP code and will serve all AutoLISP users—both novice and expert.

This chapter will cover the following topics:

- A look at the Visual LISP interface
- Exploring the Console window
- Exploring the Text editor
- Advanced programming tools
- Working with advanced programming tools
- Working with AutoCAD's database

 Note: The term "AutoLISP" should be taken to mean the AutoLISP programming language. Visual LISP is a development environment within which programs written in AutoLISP (the language) are developed, debugged, and optionally compiled into Object ARX applications. Visual LISP provides its own AutoLISP evaluator that, in effect, replaces the pre-AutoCAD 2000 AutoLISP evaluator. In this chapter, as in the AutoCAD community in general, the term "AutoLISP" is still valid and refers to the programming language used by Visual LISP.

A LOOK AT THE VISUAL LISP INTERFACE

Among its many features, Visual LISP offers an integrated set of tools and facilities specifically intended to make the writing and debugging of AutoLISP code easier. Most of these new features are visually oriented and are represented as integral components of the Visual LIPS interface. These features include the following:

- Color-coded source display

- Source syntax checker

- Autoformat and Smart Indent

- Parentheses matching

- Direct LISP function evaluation

- Source-level debugging features

The interface itself consists of an application window and several component windows. Like any Windows application, Visual LISP offers toolbars, menus, and an online Help system. The various windows can be moved, resized, and minimized. The default Visual LISP application is shown in Figure 9–1.

 Note: Although Visual LISP has the appearance of a typical Windows application with a set of windows, toolbars, and a menu bar, it cannot run independently of AutoCAD. To work in Visual LISP, you must have AutoCAD running.

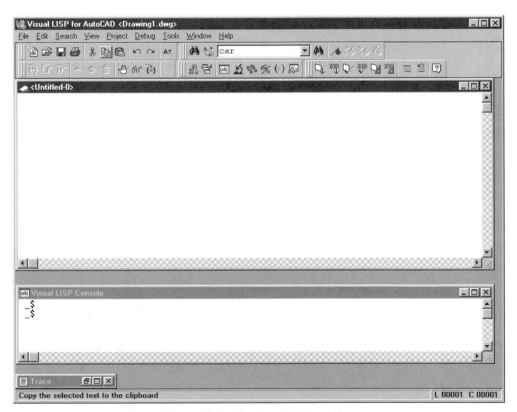

Figure 9–1 *The Visual LISP interface.*

In the following example you will start Visual LISP and explore some of its interface elements.

STARTING VISUAL LISP

1. Start AutoCAD 2000. From the Tools menu, select AutoLISP, then Visual LISP Editor. The Visual LISP window appears as shown in Figure 9–1. You can also start Visual LISP by typing **VLIDE** or **VLISP** and pressing Enter at the command prompt.

2. If your Visual LISP does not display the Text Editor window, open it by selecting New File from the File menu. You can also start a new file in the Text Editor window by typing Ctrl + N or by selecting the New tool from the Standard toolbar.

3. With Visual LISP open, use standard Windows methods to move and resize the Text Editor or Console windows. Note that the window elements within the Visual LISP interface behave as normal Windows elements.

4. From the Visual LISP menu bar, choose View to display the contents of the View menu. This menu lists the various Visual LISP interface windows.

5. From the View menu, choose Trace Stack to display the Trace Stack window. (This window may already be open.) This is one of several Visual LISP windows used in debugging Visual LISP code. Close this window by selecting the "X" button in the upper right-hand corner.

6. If you are not continuing with the next example, close Visual LISP: from the File menu, choose Exit, or select the Close button in the upper right-hand corner of the Visual LISP application window.

As you can see, the Visual LISP application window has many of the features and characteristics as a standard Windows application. Sub-windows may be opened, moved, sized, and closed. In addition, the five Visual LISP toolbars can be displayed or hidden and placed in either a docked or floated location. To control some of these features, you can use the Toolbars option on the View menu. The features of the individual Visual LISP interface elements allow you a large degree of flexibility in configuring the appearance of the Visual LISP interface.

 Note: To appreciate and utilize many of the features of the Visual LISP programming environment, you need to be familiar with the AutoLISP programming language. Most of the discussion found in this chapter assumes such knowledge. To learn more about AutoLISP, see Chapter 7, Introduction to AutoLISP Programming.

EXPLORING THE CONSOLE WINDOW

The Visual LISP Console window (shown in Figure 9–2) allows you to enter and run AutoLISP expressions and see the results immediately. In many ways it serves as a "scratch pad" for testing or verifying AutoLISP code. In this respect it serves much the same as the AutoCAD command window, where you can also enter AutoLISP code. However, the Console window offers many advantages over typing AutoLISP input at AutoCAD's command prompt. The following is a summary of what some of the important features of the Console window allow you to do:

- Enter multiple AutoLISP expressions on multiple lines. Press Ctrl+Enter to continue typing without forcing an evaluation.

- Evaluate multiple expressions simultaneously.

- Return to previously entered expressions using the Tab key. The Console window retains a history of all entered expressions. You can "scroll" backward using the Tab key, and scroll forward with Shift+Tab.

- Perform an associative search through the input history. For example, typing **foreach** and then pressing the Tab key will find and retrieve the last expression in the history that began with "foreach." You can reverse the direction of the search with Ctrl+Tab.

- Press Esc to clear any text at the current Console prompt.

- Keep a log record of all Console activity in a log file.

- Toggle to AutoCAD mode via a right-click menu to transfer Console input to AutoCAD's command prompt.

- Display the value of an AutoLISP variable by merely typing the variable and pressing Enter.

- Cut and paste text from the Console to the Text Editor window.

- Perform automatic color-coding. For example, AutoLISP functions have the color blue, parentheses are red, strings are magenta, etc. Matching parentheses are highlighted as they are entered.

Figure 9–2 *The Visual LISP Console window.*

The Console window is also where Visual LISP displays messages, including diagnostic messages, from code entered in the Text Editor. Together, these and several other features make working with AutoLISP code in the Visual LISP Console window both easier and less error prone. Combined with the Visual LISP Text Editor window, the Console provides a powerful development tool for developing both simple and complex AutoLISP code. In the following example some of the features of the Console windows will be demonstrated.

 Note: In many of the following examples you will enter AutoLISP expressions into either the Visual LISP Console or Text Editor windows. In these examples, this data will be shown in bold-faced type.

In the following example you will start to explore some of the interface elements of Visual LISP.

EXPLORING THE CONSOLE WINDOW

1. Continue from the previous example or start Visual LISP. Make the Console window current by clicking anywhere in it or by choosing the Window menu and then selecting Visual LISP Console.

2. As you type the following code into the Console window, note the color-coding for each type of element. Do not type the leading underscore or $ character; they form the Console window's prompt. Enter only the text shown; do not yet type matching right parentheses.

   ```
   _$(setq xyz (* (+ (/ 6 2.5
   ```

   Note that AutoLISP functions are blue, integers are green, real numbers are teal, and parentheses are red.

3. As you complete the expression in step 2 by typing the required closing, or right, parentheses, note that as each right parenthesis is typed, its matching opening, or left, parenthesis is momentarily highlighted with the cursor bar. After you type the four closing parentheses, do *not* press the Enter key. The completed expression will be as follows:

   ```
   _$(setq xyz (* (+ (/ 6 2.5))))
   ```

4. Continue typing *on the same line* and enter the following AutoLISP expression. Again, note the highlighting of parentheses pairs as you enter the matching closing parenthesis. (Do not press Enter yet.)

   ```
   _$(strcat "@" (rtos xyz) "<" (rtos 45 2 2))
   ```

5. Now press Enter. Note that the results of evaluating both expressions are returned. Also note that Visual LISP assigns the color green to string elements. The Console should now display the following:

```
2.4
"@2.4000<45.00"
_$
```

6. Cycle back through the Console input history buffer by pressing the Tab key once, then once again. Cycle forward by pressing the Shift+Tab keys once, then once again.

7. If you are continuing on to the next example, leave Visual LISP open.

This example demonstrates some of the features that allow the Console window to be used as a powerful adjunct to writing and testing your AutoLISP code.

EXPLORING THE TEXT EDITOR

In a very real sense, the Visual LISP Text Editor is the central and most important component of the Visual LISP environment. While the Visual LISP Console window serves you well for entering short AutoLISP expressions, you will use Visual LISP's Text Editor for extended AutoLISP code that you want to save to a file for later use. If you have written AutoLISP code in the past, you have undoubtedly used some form of a text editor, even if it was only the Windows Notepad included with the Windows and Windows NT operating systems. Notepad, and other generic text editors, offer basic text editing functions and the ability to save text to a file but they lack any tools intended for use in a programming context.

Visual LISP's built-in Text Editor is designed to meet a programmer's needs and has a number of features and tools specifically intended to support and enhance AutoLISP programming. These features include the following:

- **Color-coding**—As you enter text into the Text Editor, it is automatically color-coded based upon its specific AutoLISP functionality. For example, all internally defined AutoLISP functions are coded in blue, parentheses are red, string elements appear in magenta, etc. This automatic color-coding enables you to quickly identify various programming elements and also serves as an effective debugging aid.

- **Parenthesis matching**—AutoLISP is notorious for its parentheses, and unmatched parentheses are one of the leading causes of errors in a program. With parentheses matching, you can track parenthesis pairs as you enter code and easily find any given parenthesis's match in completed code.

- **Formatting**—As you enter AutoLISP code, the Text Editor is "intelligent" enough to apply automatically appropriate indentation, making your code easier to read. After code has been entered (or imported), you can apply formatting using any of several formatting styles.

- **Syntax checking**—You can check all or any portion of typed code for syntax errors. The specific error and the offending expression are printed in a separate window.

- **Expression checking**—You can test portions of entered code without leaving the editor.

- **File searching**—You can search for a word or expression in several files with a single command.

- **Immediate help**—You can highlight any function and press Ctrl+F1 or the Help button on the Tools toolbar to display the Help page for that function.

- **Advanced debugging**—There are a number of advanced debugging features, such as breakpoint insertion, variable watching, stack tracing, animation, and AutoCAD entity viewing.

All of these features are intended to increase the efficiency with which programming code can be entered and debugged. Tools such as parentheses checking and autoformatting can be used and appreciated by even the novice AutoLISP programmer working with relatively simple coding, while the more advanced debugging tools enable the detection and correction of errors in more advanced Visual LISP projects. Together, the features designed into the Visual LISP Text Editor make the writing of AutoLISP code and the development of Visual LISP projects easier.

In the following example you will explore some of the features that make the Visual LISP Text Editor such a useful tool in AutoLISP programming.

 Note: As with all of the AutoLISP code used in this chapter's examples, the code in the following example serves no practical purpose other than to demonstrate Visual LISP functionality. Some knowledge of AutoLISP is assumed.

EXPLORING THE VISUAL LISP TEXT EDITOR

1. With AutoCAD running, start Visual LISP by typing VLIDE and pressing Enter at the command prompt. Depending upon how you last exited Visual LISP, there may be a file opened in the Text Editor.

2. Open a new Visual LISP file: from the File menu, choose New File, or select New File from the Standard toolbar. Your current AutoCAD/Visual LISP interface should resemble Figure 9–3.

3. As you type the following code into the Text Editor, observe the automatic color-coding. Note that when you type a right parenthesis, the matching left parenthesis is momentarily identified by a cursor bar. Also note that when you press Enter at the end of the first line, the second line is automatically indented

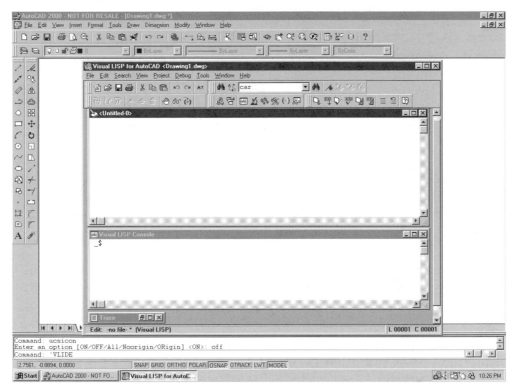

Figure 9–3 *The Visual LISP interface with a new file opened.*

because this line follows a "(defun...)" expression. Carefully enter the following code as shown, disregarding the syntax error:

```
(defun foo2 (a b)

(sqrt a b)

(print b))
```

4.   The code for the (foo2) function contains a syntax error. To see how the Visual LISP syntax checker works, from the Tools menu, select Check Text in Editor; or press Ctrl+Alt+C. The *sqrt* function contains too many arguments. Visual LISP displays the error message, as shown in Figure 9–4, identifying both the error and the offending expression.

5.   Correct the error by removing one argument from the *sqrt* function call. Note that the Text Editor behaves as a standard text editor when editing entered text. Correct the code to appear as follows (note the change to the last line):

```
(defun foo2 (a b)

(sqrt a)

(print (+ a b))
```

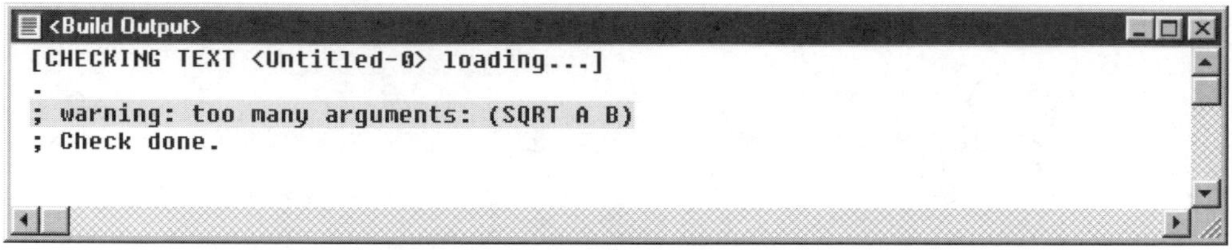

```
■ <Build Output>                                                    _ □ ✕
 [CHECKING TEXT <Untitled-0> loading...]
 ·
 ; warning: too many arguments: (SQRT A B)
 ; Check done.
```

Figure 9–4 *The syntax checker displays the error and the offending expression.*

6. Attempt to load this function into memory: with the Text Editor window active, from the Tools menu, choose Load Text in Editor; or press Ctrl+Alt+E. A load error is reported in the Console window as shown in Figure 9–5. A "malformed list" error usually means the code has a matching parenthesis error.

7. To find the missing closing parenthesis, begin checking opening parentheses. With the Text Editor active, place the cursor immediately in front of the parenthesis at "(print..." and double-click. Visual LISP highlights the opening and closing parentheses (if any) and all expressions in between. In this case the entire "(print...)" expression and its nested expression are properly formed. See Figure 9–6.

8. Check the entire function for balanced parentheses. Place the Text Editor cursor immediately in front of the first opening parenthesis for the entire function "(defun foo2..." and double-click. Because no matching parenthesis is found, the entire function requires a final closing parenthesis.

9. Use Visual LISP's formatting tool to supply the missing parenthesis: from the Tools menu, choose Format code in Editor; or press Ctrl+Alt+F. Visual LISP displays the dialog box shown in Figure 9–7. Choose Yes to have Visual LISP automatically place the missing parenthesis. Choose Yes when Visual LISP informs you that one closing bracket (parenthesis) was added to your code.

    After formatting, the code will appear as shown in Figure 9–8. The Visual LISP formatting tool corrects and formats the code in one step.

10. Attempt to reload the *foo2* function: with the Text Editor window active, from the Tools menu, choose Load Text in Editor; or press Ctrl+Alt+E. The message in the Console window indicates that the function was loaded successfully, as shown in Figure 9–8.

11. Test the *foo2* function by typing the following in the Console window and then pressing Enter:

    ```
    (foo2 16 3)
    ```

    The function returns and prints the evaluation of its two arguments, the number 19.

12. Check the current value of the *foo2* function's two variables, *a* and *b*: At the Console window's _$ prompt, type the two variable names separated by at

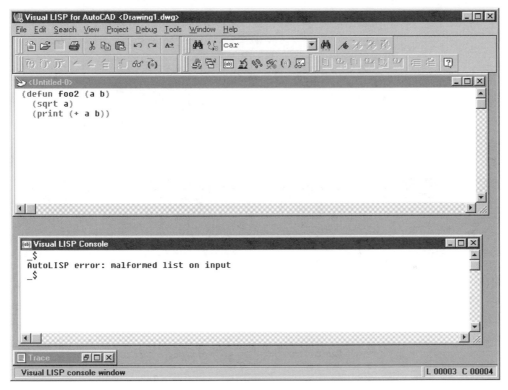

Figure 9–5 *Loading errors are printed in the Console window.*

Figure 9–6 *Checking for matching parentheses.*

least one space, then press Enter. Visual LISP reports that both variables have lost their local *foo2* values and now have a value of nil, as shown in Figure 9–9.

13. If you are continuing on to the next example, keep Visual LISP open.

In the preceding example you gained skills in using some of Visual LISP's programming tools. You saw how the Text Editor functions as a "LISP-aware" editor and

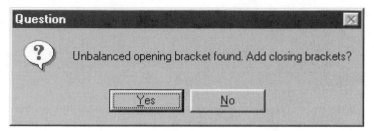

Figure 9–7 *Formatting adds the missing parenthesis.*

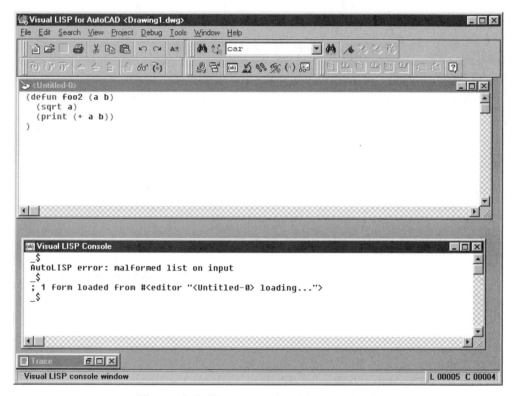

Figure 9–8 *The corrected and formatted code.*

how it interacts with the Console window. Although they were not shown in this example, both the Text Editor and Console windows have useful right-click menus that allow you to bypass the pull-down menus to initiate many of the functions you use frequently while working in these two windows. Keep in mind that the Text Editor, in addition to supporting a large number of LISP-specific functions, possesses virtually all the standard file editor features, such as cut and paste, text find, multiple file open and save, undo, etc. Most of your Visual LISP development and

Figure 9–9 *Checking the value of a variable in Visual LISP.*

debugging activities will utilize the Text Editor and the tools discussed in this section. In the next section you will investigate some of the more sophisticated abilities available to you in Visual LISP.

ADVANCED PROGRAMMING TOOLS

As some of the preceding examples have shown, programming tools such as color-coding, parentheses, syntax checking, and auto formatting are very helpful in writing and debugging AutoLISP code. Chances are, these tools will be the ones you use most often, because they address the vast majority of LISP programming concerns. Visual LISP, however, also offers a set of more advanced programming tools. These include the following:

- **Break loop mode**—You can halt program execution at user-specified points to inspect and modify the value of programming objects such as AutoLISP variables, expressions, and functions.

- **Animation**—You can watch as Visual LISP steps through your program code and evaluates each expression. Each expression is highlighted as it is evaluated.

- **Inspect**—You can obtain detailed information about an AutoLISP object in a separate Inspect dialog box. Nested objects (such as association lists) can be expanded or recursively examined down to an atomic (e.g., symbol, number) level.

- **Watch window**—You can watch the value of user-identified variables during program execution.

- **Trace facility**—This mimics the standard AutoLISP trace function that allows the printing of calls and returned values of traced functions in a devoted Trace window.

- **Trace stack**—You can view the contents of AutoLISP's call stack. Visual LISP records the history of called functions as they are executed.

Although an explanation of the functionality and ramifications of these advanced tools is beyond the scope of this chapter, the remaining chapter examples offer an introduction and brief demonstration of the power and flexibility of several of these advanced features.

WORKING WITH ADVANCED PROGRAMMING TOOLS

The following example demonstrates several of Visual LISP's advanced tools. The Watch window facility, the halt execution (Break Point) tool and the animation and formatting tools can be used to develop and debug written code. The Animation feature not only steps through code, highlighting each expression as it is encountered, it also serves to slow execution down so that various program features such as programming loops can be observed. It is often helpful to use the animation feature in conjunction with a Watch window so that you can observe the value of selected variables as they change during program execution. The Autoformat feature is not only a timesaving tool for use while entering AutoLISP code, it also serves to visually present program code in a form in which program flow is more immediately apparent.

 Note: The program used in the following example serves no practical purpose other than to demonstrate some of the advanced Visual LISP programming features. A basic knowledge of AutoLISP is assumed.

EXPLORING ADVANCED VISUAL LISP PROGRAMMING AIDS

1. In AutoCAD, open the Chap-32.dwg file from the accompanying CD. (This drawing currently contains a small donut in the lower-left corner of the opening view.) If necessary, start Visual LISP by typing **VLIDE** and pressing Enter at the command prompt. Depending upon how you last exited Visual LISP, there may be a file opened in the Text Editor.

2. From the Visual LISP File menu, choose Open File. In the Open File to Edit/ View dialog box, navigate to this book's CD and select the ANIMATE.LSP file from the chapter 9 directory and open it in the Text Editor. Your Text Editor window should now resemble Figure 9–10.

    ANIMATE.LSP defines a function that draws seven circles of radius 0.5 units, 1.0 unit apart vertically. Each circle is drawn in a different color. The program ends by erasing all seven circles, one at a time. The program, as written, requires no user input. The heart of the program consists of two repeat loops—one to draw the circles, change the color, and increment the center point. The second repeat loop merely repeats the ERASE command.

3. First, use the Visual LISP formatting feature to format the code into a more readable form. With the Text Editor window active, from the Tools menu, choose Format code in Editor. The code will now resemble that shown in Figure 9–11.

    Reformatting the code made the intent of the C:ANIMATE function clearer. The repeat loops are clearer, for example.

```
animate.lsp                                              _ □ x

(Defun C:animate (/ pt1 cl)
  (setq pt1 '(1 1 0))
  (setq cl 1)
  (command "zoom" "L" '(0 0 0) 10)
  (repeat 7
  (command "color" cl)
  (command "circle" pt1 0.5)
  (setq cl (1+ cl))
  (setq pt1 (mapcar '+ pt1 '(0 1 0)))
  )
  ;;erase circles
  (repeat 7
  (command "erase" "L" "")
  )
)
;;;;;;;;;;;;;;;END
```

Figure 9–10 *Loading ANIMATE.LSP into the Text Editor.*

```
animate.lsp                                              _ □ x

(Defun C:animate (/ pt1 cl)
  (setq pt1 '(1 1 0))
  (setq cl 1)
  (command "zoom" "L" '(0 0 0) 10)
  (repeat 7
    (command "color" cl)
    (command "circle" pt1 0.5)
    (setq cl (1+ cl))
    (setq pt1 (mapcar '+ pt1 '(0 1 0)))
  )
  ;;erase circles
  (repeat 7
    (command "erase" "L" "")
  )
)
;;;;;;;;;;;;;;;END
```

Figure 9–11 *ANIMATE.LSP reformatted in a more readable form.*

 Note: Both the color-coding and formatting features are customizable. This chapter uses the Visual LISP default settings for both features. Consult the Visual LISP documentation or online Help for details about customizing these features.

4. Next load the C:ANIMATE function into memory. From the Tools menu, choose Load Text in Editor. The result of the loading attempt is reported in the Console window (see Figure 9–12).

5. From the Debug menu, choose Animate. This will activate the Animation feature.

Figure 9–12 *The result of loading program code is shown in the Console window.*

 Note: You can control the speed of the animation. From the Tools menu, choose Environment Options, then General Options. In the General Options dialog box, choose the Diagnostic tab. In the Animation Delay input box, enter the delay, in milliseconds, for each animation step. A delay of 500 milliseconds is usually long enough to allow the progress of the evaluation to be seen clearly.

6. Next, set a Watch window to observe the value of the program's variables *cl* and *pt1*. On the first line of the function definition, highlight the variable *cl*. Then from the Debug menu, choose Add Watch. Visual LISP displays the Watch window, which displays the current value (nil) of variable *cl*. Add the program's variable *pt1* by adding it to the Watch window using the Add Watch tool, as shown in Figure 9–13, and typing **pt1** in the Add Watch dialog box. The Watch window will now resemble Figure 9–14.

7. If necessary, resize and move the Visual LISP application window until the small donut in the lower-left corner of the drawing view is visible (see Figure 9–15).

 Note: You may want to rerun C:ANIMATE to watch different aspects of the animation. With the Console window active, you can use the Tab key to recall the C:ANIMATE command. Remember that you can also slow down or speed up the animation as described in the previous Note.

8. Click in the Console window to make it active. Run the C:ANIMATE function by typing the following at the _$ prompt in the Console window and then pressing Enter:

```
(c:animate)
```

Watch the C:ANIMATE program run. Note that the program executes one expression at a time as the Animation feature highlights and pauses at each expression. As line 7 is executed, AutoCAD draws a circle. In the next line of code, the variable *cl* is incremented by 1. This is shown in the Watch window. In the next line, the variable *pt1* is incremented, as is also shown in the Watch window.

Figure 9–13 *Adding a variable to the Watch window.*

```
(Defun C:animate (/ pt1 cl)
  (setq pt1 '(1 1 0))
  (setq cl 1)
  (command "zoom" "L" '(0 0 0) 10)
  (repeat 7
    (command "color" cl)
    (command "circle" pt1 0.5)
    (setq cl (1+ cl))
    (setq pt1 (mapcar '+ pt1 '(0 1 0)))
  )
;;erase circles
(repeat 7
  (command "erase" "L" "")
  )
)
;;;;;;;;;;;;;;END
```

Figure 9–14 *Adding a second variable to the Watch window.*

9. Watch the values of the two variables change as the repeat loop is executed.

10. After line 7 passes, the repeat loop is complete and the second repeat erases the circles one at a time. Note also that because the two variables are "local" to the function, they return to a value of nil upon completion of the function.

11. If you are continuing to the next example, leave the CHAP-10.DWG and Visual LISP open. Close the Watch window.

In the preceding example you used some of the more advanced programming tools available in Visual LISP. Break points can be very useful in helping to find the specific point in a program where an error occurs. The Watch function records the value of variables—even local variables—as the program is executed, and it allows these variables to be inspected in the context of the program. The Animation function not only "forces" a series of mini-break points, it also allows you to watch the program progress visually.

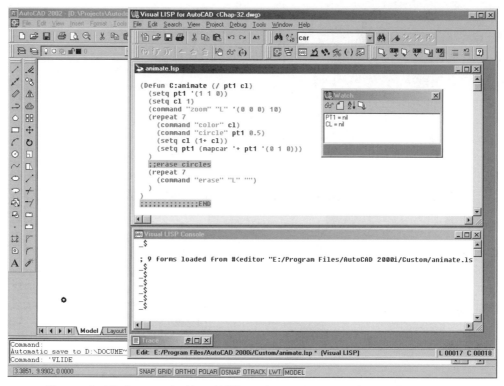

Figure 9–15 *Arrange the Visual LISP window to show the left side of the drawing.*

WORKING WITH AUTOCAD'S DATABASE

Visual LISP provides tools for viewing objects (a.k.a. entities) in the AutoCAD database. You can select an entity and inspect the object's raw data. This is the same capability that AutoLISP provides through its *entget* function. With the Inspect tools in Visual LISP you can quickly display the data of one or more objects. You can also view the various AutoCAD Symbol Tables, such as the tables for layers, linetypes, styles, user coordinate systems, etc. Block and extended data tables are also accessible through the Inspect tool.

In the following example you will use Visual LISP's Inspect facility to view AutoCAD object data.

 Note: It is not the intent of this chapter to discuss drawing database structure or manipulation. The following example is intended only to demonstrate how AutoCAD object data can be accessed from within Visual LISP.

USING VISUAL LISP TO VIEW OBJECT DATA

1. Continue from the preceding example or start AutoCAD and open the CHAP-10.DWG file from the accompanying CD. This drawing contains a small donut in the lower-left corner of the drawing limits. If necessary, start Visual LISP by typing **VLIDE** and pressing Enter at the command prompt. Depending upon how you last exited Visual LISP, there may be a file opened in the Text Editor.

2. Switch to AutoCAD by clicking anywhere in the AutoCAD window or, from the Visual LISP Window menu, by choosing Activate AutoCAD. If necessary, perform a ZOOM/All and pan to place the donut in the lower-left corner.

3. Use the Alt+Tab combination to return to the Visual LISP window.

4. From the Visual LISP View menu, choose Browse Drawing Database, then Browse Selection. Visual LISP is minimized.

5. In AutoCAD, at the "Select objects:" prompt, select the donut object and press Enter. The Inspect: PICKSET window appears in Visual LISP as shown in Figure 9–19.

6. Double-click on the entity name to display additional data. Right-click on the entity name again to display a context menu. From the context menu, choose Inspect Raw Data. Visual LISP displays the expanded association list for the donut object as shown in Figure 9–16.

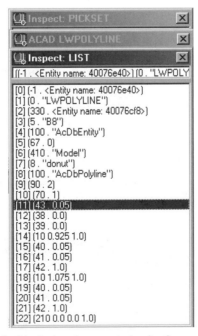

Figure 9–16 *Displaying the association list for a selected object.*

7. Close the three Inspect dialog boxes that were displayed in steps 5 and 6.

8. From the View menu, choose Browse Drawing Database, then Browse Tables. Visual LISP displays the Drawing Tables dialog box as shown in Figure 9–17.

9. Double-click on the <Layers> entry to display the layers data tables. Double-click on the <donut> layer entry. Visual LISP displays the AutoCAD Table Entry dialog box, as shown in Figure 9–18.

10. Now double-click on the {raw-data} entry to display the expanded raw data for the "star" layer. Visual LISP displays the Inspect: List dialog box containing the association list for the "donut" layer data as shown in Figure 9–20.

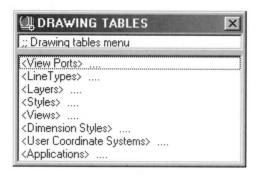

Figure 9–17 *Displaying AutoCAD tables.*

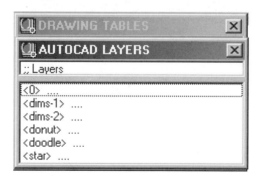

Figure 9–18 *Displaying layer table data.*

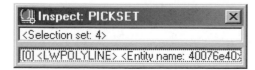

Figure 9–19 *Displaying the entity name of a selected drawing object.*

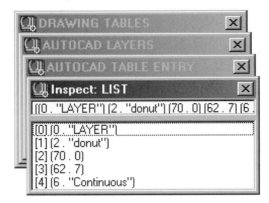

Figure 9–20 *Displaying a layer association list.*

SUMMARY

Visual LISP is an integrated programming environment within which AutoLISP programs are developed and debugged. It includes a full-featured, AutoLISP-aware text editor and a palette of tools intended to make the writing and debugging of AutoLISP programs easier and faster. This chapter gave you a brief tour of some of these facilities and tools. Several of the important features of Visual LISP fall beyond the intent of this chapter, but you should now feel comfortable experimenting within the Visual LISP development environment.

CHAPTER 10

Introduction to VBA

In this chapter Visual Basic for Applications (VBA) for AutoCAD is explored, from loading and running project macros through building simple ones of your own. Upon completion of the chapter you will have been exposed to the development environment for VBA and had a chance to see how the AutoCAD object model works. Along the way topics such as creating and editing AutoCAD objects will be covered, as well as table access, system variable access, input and output, and the preferences object.

Using one chapter to cover so many subjects necessitates keeping the subject matter concise and the coding examples to a minimum. In order to expose you to the coding, a sample project is provided in its complete form (and comments in the code are included). The coding section of this chapter will not show code for a given subject except when it is of special note. You will learn to use the step-by-step debugging feature to walk through the sample code as it runs and read the comments as you see how it works. The coding paragraphs here will give you the steps needed to start the routine and then discuss key issues to watch out for as you program your own routine of the same type. The provided code in the sample project is a good starting point for creating your own custom routines.

The first third of this chapter is devoted to conceptual issues that both programmers and CAD managers should be aware of when tackling using VBA to customize their environments. This includes the programming environment itself and how it works.

The VBAIDE, while similar to the VLIDE for Visual LISP, is different enough that you should at least skim the material even if you already feel comfortable with the Visual LISP IDE (Integrated Developer Environment).

Some general terms need to be defined to help you understand the concepts as you read this chapter. Table 10–1 contains some very short definitions of the terms that will get you started.

LOADING VBA PROJECTS

The VBA project is the receptacle for your macro's code. When Autodesk fully implemented the VBA language starting in their Release 2000 environment, they

Table 10–1: *General Terms*

Term	Definition
Project	An object containing everything relating to your program
Module	An object inside a project that contains everything relating to your code
Procedure	An object inside a module that contains your code
Object	A generic container for methods (code), properties (data), and events (code)
Method	A set of code inside an object that acts upon the object's properties (data)
Property	A storage container (variable) inside an object used to describe some aspect of the object
Events	A special kind of procedure containing code that is triggered by an event in the drawing
Dimension	Declares an object's existence
Object model	AutoCAD objects
Components	VBAIDE objects
Controls	Form objects
VBAIDE	The programming environment for VBA programming inside AutoCAD
Parameter	Data being passed into a procedure
Visual Basic Editor	A generic term for the VBAIDE

chose to give you the choice of saving one project in each drawing or saving a project to the hard drive as a .dvb file. The ability to load multiple projects into a drawing was also enabled starting in Release 2000. These two features give you many options for managing your projects and macros.

EMBEDDING A PROJECT IN A DRAWING

The advantage of embedding a project into a drawing is that its macros are available immediately when you open the drawing, without you having to load them manually or automate the loading of them. The disadvantage of this method is that each drawing must have the project embedded into it or you might call a macro that is not loaded. There are also serious maintenance issues when you want to change a macro, because you would have to change it in every drawing in which the project is embedded.

A drawing can contain only one embedded project at a time. If a drawing already contains an embedded project, you must extract (remove) it before a different project can be embedded into the drawing.

Embedding a project in a drawing is placing a copy of the project in the drawing database. The project is then loaded or unloaded whenever the drawing containing it is opened or closed.

The following steps will embed a VBA project in a drawing.

EMBEDDING A VBA PROJECT IN A DRAWING

1. Select Tools>Macro>VBA Manager from the pull-down menus or type **vbaman** at the command line.

2. Select the VBA project you want to embed in the drawing. If no project is loaded, choose Load to load one.

3. Choose Embed.

4. Choose Close to close the VBA Manager.

5. Save the drawing.

 Note: You must save the drawing to complete the embedding process. Follow the same steps to extract (remove) an embedded project from a drawing, but at step 3 choose Extract instead of Embed.

LOADING AN EXTERNAL PROJECT IN THE DRAWING

The advantage to having your projects as external files is that any project can be loaded into any drawing, and there is no overhead added to the drawing (even though the overhead otherwise would be minimal). The disadvantage is that you must load the project into AutoCAD's memory before you can execute your code, and you must organize the files on the hard drive.

Autodesk has given you many ways to load one or more VBA projects into your drawing environment.

USING THE INTERFACE TO LOAD PROJECTS

Using VBALOAD or VBAMAN (the load button) dialog boxes, the dialog interface allows the operator to select projects from the folders and load them. VBAMAN is a manager dialog box that is designed to be a central processing point for loading projects, running macros, and launching the VBAIDE to work on projects. These commands can be typed at the command line or selected from the Macros menu item under the Tools pull-down menu.

The following steps will load a project using the Load Projects (VBALOAD) dialog box.

USING THE LOAD PROJECTS (VBALOAD) DIALOG BOX

1. Select Tools>Macro>Load Project from the pull-down menus or type **vbaload** at the command line.

2. Use the dialog box's Directory list to locate your project.

3. Choose Open to load the project.

The following steps will load a project using the VBA Manager (VBAMAN) dialog box.

USING THE VBA MANAGER (VBAMAN) DIALOG BOX

1. Select Tools>Macro>VBA Manager from the pull-down menus.

2. Choose Load.

3. Use the Load dialog box's Directory list to locate your project.

4. Choose Open to load the project.

USING THE DEFAULT ACAD PROJECT

When VBA is loaded into memory during a drawing's opening, it will look in the AutoCAD application's directory for a project named acad.dvb. The acad.dvb project will be automatically loaded as the default project when it is found. You can put your code into this file and it will automatically be ready to run when the drawing opens. You can save or rename any project as acad.dvb.

AUTOMATING PROJECT LOADING USING AUTOLISP

You can use AutoLISP to load a VBA project. The command is *–vbaload*, and you must supply the path and project name. You can use the AutoLISP *findfile* function to locate your project when you have the project's location in the AutoCAD search path. The syntax for loading a project called My Project that is located in the directory path C:\My Projects, which is not located in the search path is as follows:

```
(command ".-vbaload" "c:\\my projects\\MyProject.dvb")
```

To use the *findfile* function when the project *is* located in a search path, the following code can be used:

```
(command ".-vbaload" (findfile "MyProject.dvb"))
```

You can place the loading code into any AutoLISP file, but to automate the loading of the preceding sample code, you must place it in the AutoLISP startup file that AutoCAD loads when opening a drawing. Placing the sample code for your project in the s::startup section of your acaddoc.lsp file will ensure that the project will be loaded each time a drawing is opened:

```
(defun S::STARTUP()
        (command ".-vbaload" (findfile "MyProject.dvb"))
)
```

 Note: You must turn off the Enable Macro Virus Protection feature for your automation to work without an interruption by this feature. (It throws up a dialog box that *filedia* and *cmddia* cannot control.) You can disable or enable this feature using the Options button on the VBARUN dialog box. Checking the box enables the feature, and removing the check disables it.

The AutoLISP command *vbarun* will also load a project when you specify the complete project name as part of the macro name. See the section Automating Running a Macro Using AutoLISP later in this chapter for the details.

AUTOMATING LOADING PROJECTS USING AUTOLISP IN SCRIPTS

The same code line that will load a project from the s::startup section of the acaddoc.lsp file can be used in a script to load your projects. This is very handy for situations in which you need to spin through a group of drawings and run your macros against each of them.

RUNNING VBA MACROS

Once your project is loaded into memory (through any of the methods discussed in the Loading VBA Projects section of this chapter), you need to be able to run the macros in it. This section discusses several methods for running your macros.

USING THE INTERFACE TO RUN MACROS

You can invoke a dialog interface to run your macros from either the command line or from the pull-down menus.

The following steps will run a project already loaded into memory using the Macros (Macros) dialog box.

USING THE MACROS DIALOG BOX

1. Select Tools>Macro>Macros from the pull-down menus, or type **vbarun**.

2. Select the macro you want to run from the Macro list.

3. Select the Run button to run it.

AUTOMATING RUNNING A MACRO USING AUTOLISP

AutoLISP provides a command called *vbarun* that allows you to run a macro without using the dialog interface. The command is very versatile because it not only can it run a macro, it can also load the project that the macro resides in when you specify it. An example of the syntax of *vbarun* in its simplest form (running a named macro already in memory) is as follows:

```
(command ".-vbarun"  "MyMacro")
```

This command will run the macro MyMacro if it is loaded into memory. Unfortunately, some programmers are slack about properly naming their projects and modules and macros, so you can end up with multiple projects with the same name in your environment, as well as modules and macros with the same name. AutoCAD keeps it straight behind the scenes, but it is all exposed to the operator. In that situation, the command will run the first macro it finds with the name MyMacro (sort of a "first come, first serve" kind of thing). This is very "not good" because it can end up with strange results when the wrong macro gets run.

It is better to specify the module name in front of the macro name to ensure that the correct macro is run. An example of the syntax of *vbarun* using the module name to qualify it is as follows:

```
(command ".-vbarun"  "MyModule.MyMacro")
```

The period between the module name and the macro name delineates the two objects. This is an example of the macro object (the child object) and the module object (the parent) that you will see over and over again as you learn more about object-oriented programming and the object model.

The best way to ensure that you are running the correct macro that is already loaded into memory is to specify both the project name and the module name before it. An example of the syntax of *vbarun* using both the project name and the module name to qualify it is as follows:

```
(command ".-vbarun"  "MyProject.MyModule.MyMacro")
```

The last example shows how to load and run your macro at the same time using the *vbarun* command. An example of the syntax of *vbarun* using the project path and file name, plus both the project name and the module name to qualify it, is as follows:

```
(command "vbarun"
 "C:\\My Projects\\My_Projects.dvb!MyProject.MyModule.MyMacro")
```

AUTOMATING RUNNING A MACRO USING AUTOLISP IN SCRIPTS

You can use any of the outlined commands inside a script to run your macros during a script's execution.

TESTING TO SEE IF YOUR MACRO IS LOADED BEFORE RUNNING IT

It is always a good practice to make sure that the macros you are about to run are indeed loaded into memory before you run them. Autodesk does not expose the macro name collection to us, but it does expose the project name collection (in a rather awkward and undocumented way). With the help of Randy Kintzley at Autodesk (supplying some sample code and illustrations), I was able to create an AutoLISP utility that will tell you if a supplied project (and therefore the macro in it) is present in memory. This is included as a bonus routine called IsProjectLoaded, which you only need to load into memory and call with your project name to see if the project is loaded before you try to run your macro inside the project.

Here is an example of how to use the routine to verify if your VBA project is loaded. Assume that the utility routine IsProjectLoaded.lsp is on your AutoCAD search path somewhere and that you have a project called MyProject with a module called MyModule in it. Also assume that you want to run a macro called MyMacro in MyProject. Then you could use the following code to load and run your macro from the s::startup LISP routine or from a script or from just any old LISP routine:

```
(if (setq filename (findfile "isprojectloaded.lsp"))
 (if (load filename)
  (progn
   (if (not (isprojectloaded "MyProject"))
    (command ".-vbaload" (findfile "MyProject.dvb"))
   )
   (if (isprojectloaded "MyProject")
    (command ".-vbarun" "MyProject.MyModule.MyMacro")
    (alert "Unable to locate and run MyProject.MyModule.MyMacro")
   )
  )
  (alert "Unable to load the file isprojectloaded.lsp")
 )
 (alert "Unable to find the file isprojectloaaded.lsp to load it.")
)
```

The only trick going on in this code is that the IsProjectLoaded.lsp AutoLISP routine is not silenced when it is being loaded (because it does not have as its last line of code the *princ* function to silence it). This enables the "if" statement to check if the routine has been successfully loaded. (The IF statement would return nil if it had not been successful.)

VBA/LISP AND THE OBJECT MODEL

So far you have seen a lot of AutoLISP code but no VBA code. It is evident in the body of examples that both AutoLISP and Visual LISP have strong ties to your VBA macros and are great for automating the loading and running of your VBA macros. This section explores the object model that AutoCAD has exposed to both the VBA and Visual LISP languages.

A QUICK INTRODUCTION TO THE OBJECT MODEL

Autodesk has opened the AutoCAD object model to both the VBA and Visual LISP languages. The object model is a series of definitions that allow programmers to access the data inside a drawing (both table definitions and entities). Experienced LISP programmers find it a little difficult to switch their programming style over to this object-oriented technique because there is a difference between linear LISP programming logic and modular object-oriented programming logic. The key is to "forget" the linear approach and think of it as more of a modular approach. Anyone programming in LISP for any length of time has already found the benefit to writing his or her code in a modular fashion. Object-oriented programming has taken that concept to the next level. The real difficulty for both novice and experienced programmers when they move to VBA seems to be just finding their code! Once you get used to using the tools supplied in the IDE, it becomes second nature to find the code, and you can focus on coding your problems away.

The object model is often explained as top-down or bottom-up (either approach works for trying to visualize the model), but it can also be viewed as objects inside objects, or boxes inside boxes. Think of the AutoCAD program itself as a container object. Inside it you can have one or multiple drawing objects opened. Inside a drawing object you can open a table object (like layers) or an entity object (like a line). All objects have at least two elements and some have three. Methods are an object's internal code, defining how it manipulates its own properties. Properties are an object's internally stored data.

Starting at the program object level, the application has methods that allow you to list what ARX applications are loaded or allow you to load VBA projects. It has properties that allow you to get the current active document or access the preferences object.

The drawing object has methods for opening, saving, and closing the active document. It has properties allowing you to access any of the tables, like layers or dimension styles or either paper or model space. It also has the third type of object element, which is called an event. Drawing events are triggered when you begin or end a command. They are a way to trap what is happening in the drawing environment, examine it, and make changes if necessary.

The line object has methods that allow you to set the line's color or linetype property. A line object has properties that allow you to define its color or linetype.

To explore the AutoCAD object model further, you should access a file called acadauto.chm in the AutoCAD help directory. This help file opens to the object model, and you can spend endless hours exploring all the AutoCAD objects and how they work. Placing a shortcut to the file on your desktop makes it easy to access the object model on short notice.

PROGRAMMING DIFFERENCES AND SIMILARITIES

Though this section focuses on LISP versus VBA and their access to the object model, it is noteworthy that each language has a "flavor" of its own. With LISP you can think of AutoLISP as the base language and Visual LISP as the extension of that language. With Visual Basic, VBA is the base language and VB is the extension of that language.

Visual LISP was designed to extend AutoLISP by supplying many new functions and opening up the object model to the LISP programmers. VBA was not designed to replace AutoLISP or Visual LISP, but to fill the gap between them and full-blown object-oriented ObjectArx. VBA has been "fully" implemented since the release of AutoCAD 2000. The online documentation for VBA is better than what you have gotten for AutoLISP in the past (although documentation is slowly getting better for the Visual LISP material). Table 10–2 illustrates the following features of interest about the four languages:

- The user base
- Memory usage by the language
- Development interface
- The language's orientation
- The language's ease of interface creation
- The language's method of evaluating

The User Base

Table 10–2 illustrates the growing user base for each language. AutoLISP is the oldest language related to AutoCAD and currently has the largest group of programmers using it to customize AutoCAD. Visual LISP is a newcomer to the customization of AutoCAD and its use is growing based upon its relationship to AutoLISP. Visual LISP is similar to AutoLISP; AutoLISP code runs along side it

Table 10–2: *Selected Features of AutoCAD's Four Programming Languages*

AutoLISP	Visual LISP	VBA	VB
Large AutoCAD-centric user base	Growing AutoCAD-centric user base	Large user base	Large user base
Runs inside AutoCAD memory pool	Runs inside AutoCAD memory pool	Runs inside AutoCAD memory pool	Runs outside AutoCAD memory pool
Lack of IDE and debugging environment	Has IDE and debugging environment	Has IDE and debugging environment	Has IDE and debugging environment
Not object oriented	Is object oriented (limited)	Is object oriented	Is object oriented
Difficult GUI creation	Difficult GUI creation	Easy GUI creation	Easy GUI creation
Procedural	Procedural with reactors	Event driven	Event driven

without any changes, and Visual LISP has object-oriented functions that AutoLISP does not have, along with reactors. VBA is about the same age as Visual LISP when it comes to its AutoCAD implementation, but it has a larger non-AutoCAD-centric programmer population from which to pull, as does VB. Because VB and VBA are both object-oriented languages, programmers who already know the languages must only learn the object model of the software for which they are programming.

Memory Usage by the Language

Programs that run inside AutoCAD's memory pool (internal process) will run faster (in most cases) than ones that run outside the memory pool (external process). This is because AutoCAD can handle all the resources for the program and does not have to translate anything into its memory registers. The exception to this rule is heavy database connectivity, where the application is off in a database doing queries and database manipulation instead of dealing with AutoCAD objects. Table 10–2 illustrates how the four languages use the AutoCAD memory pool. Notice that VB runs outside AutoCAD, which is usually where the heavy database applications are developed. The use of VB for heavy database applications is also due in part to the limitation of the other languages with database connectivity.

Development Interface

The IDE is a tool that all the languages except AutoLISP have. It can be argued that AutoLISP also has it, because you can develop AutoLISP code inside the

Visual LISP IDE. The IDEs in all the languages work similarly and are easy to learn. Features such as interactive debugging and easy form development (in the VB and VBA IDE) make the IDE a great environment in which to develop.

The Language's Orientation

This is how a programming language accesses data from AutoCAD. AutoLISP must use the list-based interface to access entity definitions. The language has a robust set of functions for manipulating AutoCAD through the command line interface. Visual LISP encompasses AutoLISP methods and extends them by granting limited access to the AutoCAD object model through ActiveX technology. VBA and VB expose the full object model that Autodesk has provided to-date. This is not as comprehensive as ObjectArx, but it is very powerful in its own right.

The Language's Ease of Interface Creation

Creating a dialog box in either AutoLISP or Visual LISP is not easy compared to the way VBA and VB allow you to create a form (read that as "dialog box"). This is because Autodesk still uses the old Prometheus dialog box language as the means for creating dialog boxes in both AutoLISP and Visual LISP, while VBA and VB have the standard Microsoft forms controls in them.

The Language's Method of Evaluating

AutoLISP is procedural or linear in nature when evaluating code. That means that it goes through the code one line at a time in a top-down sequence, branching off to procedures and coming back to the original procedure when it is done to finish the top-down run. Visual LISP works the same way, except that it also has reactor objects, which can be created and run in the background. These objects wait for something to trigger them and then they run the code in a top-down fashion. This is similar to the event-driven code that VBA and VB use. Event-driven code responds to registered events that an object has programmed into it. This is best shown on forms in all the languages. When you select a button on a form, the click event code is run.

THE VISUAL BASIC FOR APPLICATIONS INTEGRATED DEVELOPMENT ENVIRONMENT (VBAIDE)

The Visual Basic for Applications Integrated Development Environment (VBAIDE) is where you create your macros and programs. Because a detailed inspection of all the features of the VBAIDE could easily take up a chapter by itself, this section will focus on the main tools you will use over and over again. The Coding Your Programs section contains exercises in which you use the techniques explained in this section to explore VBA code and learn more about it. The environment follows the rules of Microsoft program interfaces, containing all the bells and whistles you find in any modern application. Exploring the environment is the best way to learn it. View this section as a primer.

You may have noticed that up until now there has been minimal mention of objects and object-oriented programming. It is important to note that while an object can represent anything, there are two different kinds of objects to discuss when speaking of programming AutoCAD (or any other application that supports ActiveX technology). There are AutoCAD objects and VBA objects. Both environments have their own object models (more on that later). All objects contain properties (data) and methods (programming) that you can code or set, and some objects (the drawing itself, forms, and form controls) even have events that can be coded. These features of objects will be explored in more detail later in this chapter in the Coding Your Programs section.

AutoCAD has internal objects (called collections), which relate to the table information stored in the drawing, and entity objects, which relate to all the different entities in a drawing. VBA has objects (called collections), which allow you to store data and manipulate it, and form objects for creating dialog boxes. The good news is that they all behave in a similar fashion. Once you learn how to manipulate a collection, you can do it for VBA data or AutoCAD data. That is the beauty of objects.

ACCESSING THE VBAIDE

The VBAIDE is only available inside the AutoCAD environment. You can access it from the Visual Basic Editor menu item found in the Macros menu under the Tools pull-down menu. You can also enter the environment by typing VBAIDE at the AutoCAD command line. Another method of accessing it is by selecting the Visual Basic Editor button on the VBA Manager dialog box. The environment gives you access to the various components that enable you to build your macros.

IDE COMPONENTS (OBJECTS)

The various components of the IDE can be moved into any configuration you desire, with the ability to turn components on and off as you need them. All the components interact with each other, providing an environment for programming that is easy to work in once you get used to it. The most used and most useful features are the following:

- Project window
- Code module window
- Properties window
- Immediate window
- Watch window

The Project Window

The Project window, shown in Figure 10–1, displays all the opened projects and their components. It is used to maneuver around in the code modules and forms.

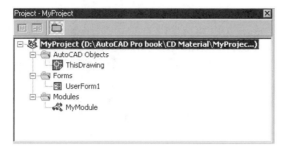

Figure 10–1 *The Project window.*

If you think of a project as an object and all of its components as objects inside that object, then the Project window is the interface that shows you their relationship to one other and allows you to maneuver around in them. Double-clicking on an item in the Project window will display the related code or form in the Code module window so you can work with the code or form. Project component objects are as follows:

- AutoCAD objects
- Forms
- Modules
- Class modules

Another technique for maneuvering around the Code module from the Project window is to right-click once over the item in the Project window and then select the View Code item from the resulting menu.

AutoCAD Objects. The AutoCAD objects component gives you access to the current drawing (ThisDrawing). Each drawing has its own AutoCAD object. Here you can use the Code module to place code, called event-based code, that will stay with the drawing and always be available. This topic is beyond the scope of a primer on the VBAIDE. However, you can experiment by placing data in the General section of the Code module or by using the Object List and Procedure List of the module to place code in the AcadDocument section. (You will learn more about the Code module later in this section.)

Forms. This component allows you to access UserForms. UserForms are special VBA objects that interact with the operator. Think of them as objects with faces on them that the operator can see. Forms contain control objects, and both forms and control objects have events that can be coded to your needs, as well as properties such as color that you can set, and methods you can program, such as how the object behaves when it is resized.

You can create a new UserForm by selecting the UserForm menu item from the Insert pull-down menu. Properties can be adjusted using the Properties window (shown later in this chapter), while the form itself can be created using the Code module and the Toolbox toolbar. (Use the View pull-down menu to enable the Toolbox toolbar.) To place controls on the form, you must activate the form by double-clicking on the UserForm you want to work with in the Forms component of the Project window. Then you can drag and drop the controls from the Toolbox toolbar. Double-clicking on a control object already on the form gives you access to the code behind the control. This is where you place your programming (the events) for the control. More on forms later.

Modules. The Module component gives you access to the general coding area of the program. This is where the bulk of your code that supports the controls and form code is placed. As you become more familiar with the various components, you will see that each one is designed to support different aspects of the structure of your program. Each component is literally just a double-click away using the Project window. There does not have to be anymore searching through endless lines of code. You select the component in which your code resides and then use the features of the Code module to work your way right to your code. Most code ends up being just a couple of clicks away from wherever you are at any given time (usually less).

Class Modules. The Class module is another component that is too complex to go into in any depth in a primer. This customization component holds the code used to create your own custom VBA objects for your program. The concept is actually quite simple. Your procedures (macros) in a Class module are the methods for the new custom VBA object, while the public variables are the properties. Once you code a custom object in this component, you can then use it anywhere in your program, just as if it were an object that Microsoft supplied with VBA. It is only available to your project, however, and only when it is loaded into memory and running.

The Code Module Window

Learn to love it because you will be working in it as long as you program VBA macros. The Code module window, shown in Figure 10–2, is the work area for all code in your program. When you double-click on any of the component items in the Project window (except the UserForms), the related code appears in this window. The actual UserForm will appear in the Code module window when you double-click on it in the Project window. To get to the code behind the form, you must double-click on the form or one of its controls in the Code module window, or use the single right-click method discussed previously in the Project windows section.

Figure 10–2 *The Code module window.*

The Code module window has many features that help you quickly find and write your code. Some of the most useful are the following:

- Object list
- Procedure list
- Coding features

The Object List and Procedure List. You can scroll through hundreds of lines of code as the programmers of old had to do, or you can use these two features to go directly to your code in just one or two picks off the lists. Which would you rather do? Do not bother to answer, because it is a rhetorical question. Of course you want to do it the easy way!

The Object list works in conjunction with the Procedure list, giving you access to your code in the Code module window. The Object list is a pop-list located in the upper-left corner of the Code module window, while the Procedure list is located in the upper-right corner of the window. Together they are the index for finding your code in the Code module (the main section of the window). The Object list is the master list. This means that you should select the object you want access to on this list first and then go to the Procedure list and select the procedure on which you want to work.

In certain situations the Object list object has only one Procedure object, so the Procedure list only has one item. In other situations the Object list object has many possible procedures (or events in the case of forms and the ACadDocument) for you to choose from, so the Procedure list has many procedure objects. The beauty of the system is that you do not have to keep track of the organization of the code, because it is done automatically by the VBAIDE. When you create a new procedure, the IDE automatically catalogs it and adds it to the list of possible procedures.

In Figure 10–2 the General object is shown in the Object list along with the Declarations object in the Procedure list. Every module object in a project has a General object in the Object list and a Declarations object in the Procedure list. This combination accesses the location where modular level global variables, constants, and declarations are set for the module. These are in effect for and available for all the procedures in the module. Setting a constant in the ThisDrawing General object makes that constant available to any of the procedures in the module. Doing the same for a constant in the UserForm module or the Code module makes the constant available to procedures in those modules.

Note: The General section begins with the statement Option Explicit. This means that you must declare all of your variables before using them or else the environment will stop you from running your program. The VBA interrupter is very forgiving in that it will take you right to the offending variable so you can see the problem and fix it. This is a great feature for avoiding the headache of wrong data types. Unlike LISP, VBA variables will not automatically switch data types as you bind data to them. When you use this feature you avoid the problem. VBA will automatically put this statement in the General section when the Require Variable Declaration box is checked in the Options dialog box under the Tools pull-down menu.

Coding Features. If you are just learning VBA do not panic yourself by thinking there is a lot of memorizing and typing involved in creating programs. You have already seen how the editor has features for helping you quickly locate your code. It also provides a fast typing mechanism for maneuvering inside the object tree of an object. When you properly dimension (*Dim*) an object type, the system uses a feature called IntelliSense that will display the properties and methods in a pop-up list for quick selection, as shown in Figure 10–3.

After typing the first character or two of the property or method you are after, the menu scrolls down to the area you are looking for. You can then select the remainder of the word with the mouse. Operators of portable computers that have the pointing device near the keyboard report that this is a really great feature. You can use keyboard strokes to select the value desired as well, and then press the Tab key when you are satisfied with the selection. This method of code input greatly enhances the speed and accuracy of the coding.

The Properties Window

The Properties window, shown in Figure 10–4, gives you access to higher-level objects' properties. With it you can set a module or project name. Forms and controls properties are also set from this window. To place an object's properties in the window, just select the object from the Projects window or, in the case of forms and controls, select them from the Code module.

```
Public Sub MyMacro()
    MsgBox "MyMacro is running."
    Dim objLine As AcadLine
    objline.la
End Sub
```

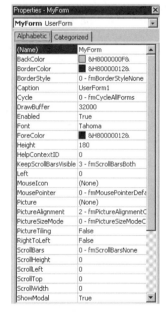

Figure 10–3 *The IntelliSense feature.*

Figure 10–4 *The Properties window.*

Setting a property during development is simply a matter of selecting the item you want to set from the right column of the window and either typing or selecting a setting from a list.

 Note: Most properties can also be set through code at run time. An example is the caption property on a form or control. For example, given a control called btnOK, you could set what the operator sees with one line of code. The syntax would be as follows:

```
btnOK.Caption = "My new caption"
```

The Immediate Window

While the Code module window may be where you do most of your work, the Immediate window, shown in Figure 10–5, is perhaps one of the most important in the development environment. Here is where you can examine variables as the program runs or reset a variable for testing purposes on the fly. You can run a new line of code in this window while the program is paused to see the effect it will have on the data. You will learn more about how to use this great feature in the Debugging Your Program section.

The Watch Window

This window, shown in Figure 10–6, is used to help debug your program. It allows you to select a variable or an object's property and watch it as you spin through the program. This is very helpful when your program is not behaving with a set of data the way you thought it would. Unlike the Immediate window, which is very flexible and dynamic, this window is static. Once you select something to view, it stays in the window for you to see, and you cannot change the data associated with it. Think of it as a snapshot of the state of the variable as it passes by the line of code in the procedure when you are running your program.

The columns in the Watch window display are shown in Table 10–3.

You will learn more about how to use this feature in the next section.

DEBUGGING YOUR PROGRAM

This is the area in which an integrated development environment is designed to shine. As you write your program you need to test it and observe that it is processing the information the way you want it to (which is not always necessarily the way you designed it). This way you can make adjustments as needed to complete the program.

Figure 10–5 *The Immediate window.*

Figure 10–6 *The Watch window.*

Table 10–3: *Watch Window Data Columns*

Heading	Description
Expression	The variable name
Value	The value the variable is set to as the procedure runs
Type	The data type of the variable
Context	Where the variable is in the code stream

The only way to do this is to step through your program one line at a time, viewing the data as you manipulate it. Debugging features that allow you to do this include the following:

- Stepping through your code one line at a time as it runs
- Setting break points to pause the program as it runs so you can examine the state of your variables
- Setting a watch on a variable
- Setting your next statement to run
- Running to your cursor position
- Printing variable data to the Immediate window as your program runs

Stepping through Your Code

The Debug pull-down menu gives you several ways to step through your code:

- Step Into
- Step Out
- Step Over
- Run to Cursor

Stepping into code (F8) moves you to the next line of code (executing the line it was on in the process). When the next line of code is a call to another procedure, the feature will take you into that procedure.

When the called procedure is tested and you do not want to step through each line of code in the procedure, you can use the Step Out feature to jump out of the procedure. This is good for situations when you accidentally move into a procedure you know is good and need to quickly bail out.

The Step Over feature is good for jumping past the next line of code. This is great for when you realize that you want to jump past a call to another procedure that you know is good.

The Run to Cursor feature is good for stepping past a whole block of code that you know works to get to a trouble area.

 Note: All of these features execute the code when they pass over it.

Break Points

You set a break point to tell the program to pause when it gets to that point in your code (so do not leave them in once you finish debugging). This feature enables you to experiment with data in your variables as the program is running or to run a small bit of code in the Immediate window without having to jump through hoops organizing your data. Placing your cursor on the desired line of code and selecting Toggle Break point from the Debug pull-down menu sets break points. You can remove an existing break point the same way. Two other shortcuts do the same thing. One is the (F9) function key, and the other is selecting a point next to the desired line on the left-hand margin of the Code module window.

Stop

This function can be placed on a line of code to pause the program when it gets to that line. It has the same effect as a break point except that it is permanent and will be saved with the project. You will want to remove or comment out any Stop functions before delivering the program to the operators.

AppActivate

This function switches you from the development environment to the AutoCAD application during testing of your program in a development session. It will be saved with the project just like a Stop function will be, so you will want to comment it out before delivering the program to the operator. The function needs to know what application to switch to, so you can use the special ThisDrawing object to tell it. The following line of code tells the function to make the AutoCAD application active by calling the *AppActivate* function and giving it the Caption property of the AutoCAD application:

```
AppActivate ThisDrawing.Application.Caption
```

 Note: The object ThisDrawing has a property associated with the application object so that you can drill into the application object from the drawing itself. This exposes the application-level properties to you.

Setting the Next Statement

This feature will actually skip over code from one point in your procedure to another. It is handy for stopping to examine a variable and then skipping some statements that you do not want to affect the data before continuing on with your code. You can set a break point and, when the program pauses, place the cursor on the line you want to execute next and then select the Set Next Statement on the Debug pull-down menu (or use the Ctrl+F9 key combination).

Debug.Print

This feature will print a string to the Immediate window. It is yet another way to view data values of variables as your programs run. This function is saved with the project and is usually commented out before delivery to the operator.

CODING YOUR PROGRAMS

Learning any modern programming language means learning the basic techniques the language uses to gather input and output, create objects, edit objects, remove objects, and interact with the object model. The rest of this chapter is dedicated to examining code and exposing you to the techniques of object-oriented programming with the AutoCAD object model.

Since there is so much code and so little room in this chapter the approach is to use the following loading and debugging techniques with the supplied project to walk through the code and learn what it is doing. All the routines in the project are heavily commented to help you understand what is happening and the following section will point out VBA programming techniques used in the routines in the sections marked as 'Code of Note'. Whenever you see a new routine move to it in the project and run it using the debugging techniques shown here.

The sample project (MyProject.dvb) provided on the media accompanying this book should be loaded into the VBA Manager of any new drawing so you can follow along and practice the techniques described.

VBAIDE SETUP

Before you start the coding, make sure the following windows are viewed in the VBAIDE:

- Project window
- Properties window
- Code module window
- Immediate window

You can set your windows up by selecting any closed windows from the View pull-down menu during the second step of the first exercise. Where you put them on the screen is up to you.

Before you explore the sample code, a few exercises are in order to get you up to speed on the debugging features so you can follow along. These techniques are the same ones you will use during testing and debugging of your programs. All the coding exercises in this chapter are based upon the debugging and testing techniques described previously in this chapter. Once the project is loaded, it is assumed to be loaded in the rest of the exercises. If the project is not loaded, you will need to reload it and run its exercises using the following steps.

RUNNING A MACRO FOR TESTING

You can use the *VBARUN* command from inside AutoCAD to start your macro for testing, but you will soon grow tired from jumping back and forth from the IDE to the AutoCAD window to start the command. A quicker technique is to run the macro from right in the IDE. Perform the following steps to run the macro that gives an example of using the Immediate window.

RUNNING A MACRO FROM WITHIN THE IDE

1. Load the project file (MyProject.dvb) using any of the techniques explained earlier in this chapter. (VBAMAN is a good one.)

2. Enter into the IDE using any of the techniques explained earlier in this chapter. (The Visual Basic Editor button on the VBA Manager dialog box works well, if you are already in the VBAMAN dialog box.)

3. Open the Modules folder (if it is not already open) and double-click on the module you want to work on (MyModule) in the Project window.

4. In the Code module window use the Procedure list to select the procedure you want to work on (ImmediateWindow_Demo).

5. With the cursor inside the procedure, select the Run Sub/UserForm button on the Standard toolbar shown in Figure 10–7 (or press the F8 key) to start the procedure. The routine will pause at the Stop function. Whereas the Run Sub/UserForm button on the Standard toolbar starts a procedure when your cursor is inside it, the Reset button (also shown in Figure 10–7) will stop the procedure. Read the following information before continuing with step 6.

Figure 10–7 *The Standard toolbar.*

Note: The first four steps are the same ones you use for working on any project. Once the project is opened in the IDE, you can skip the step that gets you there (unless you need to load another project). While this example works on an existing project, the techniques are the same for creating your own project. The only real difference is that you will not need to open the project the first time (step 1); instead you will insert the new modules and procedures and then use the Save feature in the IDE to save your new project.

The Immediate Window Demo routine was designed to show you how to print information to the Immediate window and give you a taste of using the window. It builds a collection, fills it with some data, and then spins through the collection, printing it to the Immediate window. The collection and accessing it will be discussed in more detail later.

Now you should still have the program paused at the Stop function. It is time to "play" with the Immediate window. To see the value bound to the *strString* variable, perform the following steps.

VIEWING THE VALUE BOUND TO THE STRSTRING VARIABLE

6. Move your cursor to the Immediate window by picking on it.

7. Type the following string in the Immediate window exactly as shown:

    ```
    ? strString
    ```

    The result should be that the string is assigned to the variable before the Stop function (MyString). Now change the information and view the change.

8. Type the following string in the Immediate window exactly as shown:

    ```
    StrString = "YourString"
    ```

9. Type the following string in the Immediate window exactly as shown:

    ```
    ? strString
    ```

    Your new assigned data (YourString) is now bound to the variable. This technique can be used to change almost any variable data before continuing the program as you test it.

10. Select the Run Sub/UserForm button again to finish running the routine.

The end result in the Immediate window should be the numbers 1, 2, and 3, each on its own line. This simple Immediate Window Demo routine has exposed you to several techniques you will use again and again as you program in the VBAIDE. Another feature to experiment with is the break point discussed earlier. The break point is equivalent to a temporary Stop that is not saved with the project.

Code of Note in the Immediate Window Routine

Coding Comments. You should always comment your code so that when you or someone else goes to maintain it at a later date, the intent of the code will be understood. VBA uses a single quotation mark (') to keep comments from being evaluated by the interpreter.

The Dim statement. This function declares a variable and its data type. It is always best to declare the exact data type for the variable. VBA uses the information to insure that you are using proper programming techniques. This routine declares a string variable, a long and a new collection for use in the routine. For more information on all the data types available, see the online Help index topic Data Types Keyword Summary.

The Stop statement. This function halts the execution of the routine without exiting it. Once the routine is paused, you can use the Immediate window to experiment with variables or other code fragments or assign a variable to the Watch window for reviewing as the program progresses. For online Help on the Stop function, place your cursor on the word Stop and hit the F1 key. Help will take you right to the section.

The Debug statement. This function's Print method is used to print information to the Immediate window.

INPUT/OUTPUT

A program that does not get input from an outside source is very inflexible and of little use in most situations. At the same time, a program that processes information does little good unless it can output the results. Input takes several forms. You can get data from a file, an external database, the drawing in which it is stored, or the operator. Output can take the form of writing to a file, an external database, the drawing, or the screen.

Output File Manipulation

Writing output to a file is straightforward using VBA. To see an example, open the MyProject.dvb file using VBAMANAGER and run the macro FileManipulation_Output.

 Note: The steps for loading and running a macro were covered in the Running a Macro for Testing section earlier in this section.

The routine has a Stop that pauses the routine so you can walk through it one line at a time. When the routine pauses, hover your cursor over the variable *intWfl* (without picking it). Another feature of the IDE is revealed when the value of the variable (zero in this case) is revealed. Next (leaving the cursor over the variable) press the F8 key twice to Step Into the line of code below the variable. The value on the hovering cursor should change to a one. This feature is great for getting a quick look at a value without using the Immediate window or the Watch window. Try the same thing over the *strAppPath* variable. It will start as an empty string and then be filled with the AutoCAD application path string after you move past the line of code.

Finish running the routine by selecting the Run Sub/UserForm button again. The result is that the file (MyProject TestFile.txt) has now been created in the application's directory. You can use any text editor such as Notepad or WordPad to open the file up and verify that the contents are the same as what the Print # statement placed in it.

Code of Note in this Routine

The Public and Private Keywords. When you declare a procedure, VBA needs to know the scope of the procedure. Scope is the availability of a procedure to other procedures. There are two types of scopes you can declare for a procedure. One is the Public keyword, which tells VBA that your procedure is available to all other procedures in your program. This includes procedures in other modules (like the UserForm module or other support modules you build). The other keyword available is the Private keyword, which tells VBA that your procedure is only available to other procedures in the module in which it resides. This system allows for maximum flexibility as you learn more about the capabilities of VBA. Because this procedure is a standalone procedure, it could have been declared as a Private one. Most of the time you will use Public in your declarations because it does not hurt anything that the procedure is available to other procedures.

The Sub and Function Statements. There are two procedure types in VBA: Sub and Function. They are basically treated the same by VBA except that Function statements can return data to the calling procedure whereas Sub statements cannot. More on Function statements later.

ThisDrawing.Application.Path. The ThisDrawing object has a property exposing the application object. You can drill into the property to the Path property of the application object to get the path of where AutoCAD resides. This is useful for finding support files or building your own data support structure under the AutoCAD application.

FreeFile Function. The FreeFile function keeps track of all the open files your program is handling and gives back the next available file handle as a data type of *long*. (The Open command needs the file handle to open the file and work with it.) This feature is great because it takes keeping track of the last file handle number you used out of your hands and automates it.

Open and Close Statements. These functions are used to open a file and close it. The VBA Open statement can handle binary files as well as text files. For a more detailed explanation of the parameters for this function, see the online Help for the

Open statement. Always remember to close an open file or else you could cause file access errors with other programs.

Print # Statement. This function writes data to a file. The function uses the file handle provided by the *FreeFile* function to know the correct file to which to write. The corresponding function for reading a line out of a file created using this function is the Input # statement.

String Concatenation Using the Ampersand (&) Symbol. The Open line of code uses string concatenation to tack the path onto the front of the file name. Some programmers mistakenly use the + sign to do this. While the environment currently supports it, using it is illegal, and the environment will eventually be adjusted to work only with numbers. When that happens, programs using the + sign to concatenate strings will be broken.

Input File Manipulation

Reading in data from a file is just as simple as and very similar to writing it out. To see an example, first open the MyProject.dvb file using VBAMANAGER (if it is not already open). Make sure you do the FileManipulation_Output exercise before this one because this exercise needs the file created by that exercise (and you will get an error if the file is missing). Run the macro FileManipulation_Input and step through the code from the Stop function using the F8 key.

The results should be the contents of the file displayed in the Immediate window.

Code of Note in this Routine

Do Loop Statement. This function will repeat a block of code while a condition is **true** or until a condition becomes **true**. It is commonly used with the *EOF* function to spin through a file and upload the data from the file into a collection for later processing. For more detailed information on the Do Loop statement, see the online Help.

EOF Function. This function returns a Boolean **True** when the end of a file is reached. It is used to break out of a Do Loop. See the online Help for more details.

SELECTING OBJECTS

Because VBA macros must have the AutoCAD drawing editor active, there is always at least one current drawing active. As mentioned earlier, the current drawing (or document) is defined as an object with the name ThisDrawing in the VBA environment. ThisDrawing is essentially the trunk of the object tree, with the application object as the root.

The object tree can be explored by referencing the reserved object variable *ThisDrawing*.

```
Public Sub Main()
   Debug.Print ThisDrawing.Name
End Sub
```

There are many ways to use the ThisDrawing object to get selected input from the drawing and operator. Some methods allow you to select geometry automatically based upon criteria (called filters) you set up, and others allow you to give control to the operator to select the geometry.

Automated Geometry Selection

Before you start the CreateEntsCollection exercise, jump over to the drawing (you can use the key combination Alt+Tab to go back and forth) and place at least one circle in model space (also place a block if you like). The routine will automatically find all circles and blocks in model space and show you a count in a dialog box at the end. This exercise is designed to introduce you to several new concepts while showing you how to automate the process of gathering objects from the file without operator interaction. Among the concepts illustrated in this exercise are the following:

- Passing parameters between subroutines
- Collections
- On Error
- For Each statement
- If statement

To begin, open the MyProject.dvb file (if it is not already open) with VBAMAN-AGER. Run the macro CreateEntsCollection and step through the code from the Stop function using the F8 key.

Code of Note in this Routine

The Function Statement. The Function statement allows you to return data to the calling procedure. The *GetAllEnts* function is designed to accept the AutoCAD space object (model or paper) and the entity name you want to collect. It then does all the work, collecting all the entities into a collection and returning that collection to the calling procedure.

The *GetAllEnts* function is an example of a portable support toolbox procedure that you can use over and over again in your programs. Toolbox routines are designed to be black boxes. They take defined data in and return defined data out. Once the routine is designed and tested it can be used over and over again. In this example the *GetAllEnts* function accepts either model or paper space and any legal entity name and always returns a collection containing any of the objects found in that space that match the entity name. It can be called from any procedure in the program and will always return a collection (either empty or with entities in it). The calling routine only needs to check if the collection has anything in it to proceed with processing the entities.

Passing Parameters between Subroutines. Rather than placing all your code in one procedure and having to scroll through thousands of lines of code for a program, you can modularize your code into subroutines. One of the key abilities of any higher-level language is the ability to pass data back and forth between routines. This ability allows you to organize your code into small, easily manageable segments. This is not a new concept. Anyone familiar with LISP is aware of modularizing your code for flexibility. VBA allows passing of most data types both ways (into and out of modules).

Passing Data In. VBA procedures must declare the passed-in variable in the parentheses as part of the subroutines declaration. The generic syntax for a subroutine procedure is as follows:

```
Public Sub ProcedureName (parameter1 as String, parameter2 as
    Collection)
```

In this code example, the first parameter passed in will be called parameter1 and will be a string data type. VBA will ensure that it is a string by checking the data type of the variable being passed. It is important that the data you pass in is of the same type declared or else VBA will stop you with an error. The second variable is declared as a collection object called parameter2. Once the program moves into this subroutine, the two parameters are available to the subroutine for processing.

An example of a called subroutine taking a parameter is as follows:

```
Public Sub ViewObjectCollection(Collection As Collection)
```

The ViewObjectCollection subroutine is being used by the main procedure to display the entity information of any entities found by the *GetAllEnts* function. It can not do its job unless the calling procedure has gathered a collection of entities and passed them in. This makes the CreateEntsCollection procedure the driving or main procedure coordinating the function that gathers the desired entities and passing them to the viewing subroutine for display. The main routine remains small and easy to read and maintain.

Passing Data Out. Only the Function procedure can pass data back to a calling procedure. You declare the passed-out data type at the end of the procedure declaration. The generic syntax for a Function procedure is as follows:

```
Public Function ProcedureName (paramerter1 as String, parameter2 as
    Collection) as Collection
```

Just like the Sub procedure illustrated earlier, the passed-in parameters are listed inside the parentheses. As VBA evaluates the code, it knows that you can pass back a collection because you declared it at the end of the procedure declaration. It will only allow you to declare a return variable if you declared the procedure as a Function.

An example of a called Function is as follows:

```
Public Function GetAllEnts(Space As Object, objName As String) As
    Collection
```

This toolbox function is used in the example to gather circles from model space and pass them back to the calling routine. It will accept any legal entity name and either space object.

Once you have declared the type of object you are returning in the procedure's declaration, you must use the procedure's name to pass the result back. The variable you pass back must match the declaration you use in the procedure declaration. The idea is to do the work in the procedure and then pass the results back. The line that passes back the data in the *GetAllEnts* function is the last line in the function, as shown here:

```
Set GetAllEnts = colObjects
```

When the data type of the object being returned is a collection, you must use the Set statement and the function's name with the equal sign to pass the temporary collection back to the calling routine.

Collections. This is the second exercise that uses a collection. Collections are storage objects that have no limits on the amount of data they can store and can contain any data of any data type. Collections can even hold other collections. A collection is the ultimate storage device in VBA (and AutoCAD). You can spin through a loaded collection to access its stored items one at a time, or you can assign a unique key to each item in the collection as you store the item (sort of an address in English) and then access the associated data through the key. See the online Help for more details on the methods the collection has to enable you to add and access data.

The New keyword is used when you create a new collection object to initiate the object at the same time you declare it. This allows you not to have to use the Set statement to initiate it. Once the collection object is initiated, you can use the Add method with no equal sign. (Truth be told, all a method on an object is is a procedure that takes your data as a parameter.)

For Each Statement. The example of this special kind of loop for spinning through collections or arrays is in the *GetAllEnts* function. Because the space object in a drawing is a collection, it can be used to spin through the space object so you can check each entity object and see if it is the desired object. When it is the desired object type, the object is added to the internal collection of the routine for passing back to the calling routine. Every For statement must end with a Next statement. Any code between the beginning and end parts of the statement is executed until the condition is no longer true on the starting For Each part of the statement.

The Generic Object. When the *GetAllEnts* function spins through the space to look at each object, it does not know what kind of object it will get next. The generic object is designed to accept any kind of object. It has some generic properties that are common to all AutoCAD objects, and you can query that information once the object is loaded to find out what kind of object is inside the generic object. An example of the declaration of a generic object is as follows:

```
Dim Ent As Object
```

If statement. This statement shows up in all three procedures. The statement evaluates the supplied data and determines when it is either true or false. When the data is found to be true, the first block of code is executed otherwise the block of code after the Else keyword is executed.

CStr Conversion Function. Conversion functions are necessary to change the data type of a variable when you have data in one form but need to process it in another form. The example in the ViewObjectCollection subroutine takes a counter number and converts it to a string to concatenate to another string for display in a message box. To read more about conversion functions, place your cursor on the *CStr* function and hit your F1 key.

Msgbox Function. This function displays a string in a special dialog box. You can control which buttons are displayed on the dialog box as well as what the title will be.

The Nothing Keyword. This keyword is used to set existing objects to nothing (thus clearing them of any data bound to them).

Call Statement vs. No Call Statement. You may have noticed in the CreateEntsCollection procedure that the ViewObjectCollection subroutine is called two different ways, once with the Call statement and once without. You may also have noticed that the passed-in parameter was enclosed in parentheses when the Call statement was used. and not enclosed when it was not used.

The Call statement's job is to transfer control to a subroutine or function. When the Call is used, the returned value from a function cannot be caught. Thus the need to be able to call a function without the Call statement arises. For more details see the online Help.

Operator Geometry Selection

The GetEnt subroutine is an exercise illustrating how to get the operator to select an object for you. It is a simple one-time-only try at getting the geometry. If the operator cancels the routine's execution or misses the geometry, the routine displays a message indicating that nothing was selected. The routine illustrates the following new techniques:

- Using one of the utility object's *Get* functions
- Using error trapping to query the status of an object
- Highlighting a selected object

You can wrap the Utility.GetEntity method in a loop, testing to make sure that the operator did select an object before you move on to process it.

Code of Note in this Routine

Using a Utility Get Function. Each drawing has a utility object that contains various functions for getting information or calculating data. The GetEnt routine uses the GetEntity method, allowing the operator to select an object.

Using Error Trapping to Query an Object. Sometimes it is difficult to determine when an object is initialized with data. The GetEnt routine uses a technique with the On Error statement to check if an object is set. After the Utility.GetEntity call has been made, the routine uses Debug.Print to try to print out the entity object's name. The Debug.Print statement will throw an error if the object it tries to print the name from is not set. The If statement is used to test the error object's number. If the number of the error object is anything but zero, then an error occurred when the statement tried to print, which means that the object was not set. Error trapping warrants a whole chapter by itself and is somewhat of an art.

Highlighting a Selected Object. When an object is selected, you might sometimes want to show the operator that it is selected while you are off processing it (just like AutoCAD does). Each AutoCAD entity object has a Highlight method that will accept a setting of True or False. When set to True, the object is highlighted on the operator's screen. When False, the object is not highlighted on the screen.

CREATING GEOMETRY

Although the data needed to create a simple piece of geometry like a line or a circle is different in most cases the process for creating them and setting their properties is about the same. The AddLine exercise is designed to show you that process while introducing you to some new techniques. In this exercise are the following:

- Model space shortcut
- The *Chr* function
- VBA constants
- Arrays
- GetPoint method
- *IsEmpty* function
- Object Update method

Code of Note in this Routine

Creating an AutoCAD Object. Most AutoCAD objects need at least one point in order to place them in the drawing. Some need a radius, a scale, or a rotation, while all can have a color or linetype set. The steps taken to place a new object in the drawing are pretty much the same every time. You need to gather all the necessary information to create the object, declare a container object to hold the new object, and then issue the proper Add method from the correct space collection. Updating the object to the screen is optional but encouraged.

Model Space Shortcut. This exercise illustrates how you can save some typing by setting up a shortcut for collections or objects that are often used. To get to objects in model space you would have to type Thisdrawing.ModelSpace each time you wanted to add, edit, or access in any way an object in model space. There are two techniques for doing this. One is the technique shown in this exercise, where you declare a model space container object and then you bind it to model space. Once you do this, you can access any of the model space objects methods or properties through the shortcut variable you defined.

The other technique not shown in this exercise is the With statement (which is explained in more detail later in this chapter).

The Chr Function. VBA provides the *Chr* function to convert a character code (usually an ASCII code) to a string character code. The example in AddLine using *Chr(10)* and *Chr(13)* creates a carriage return–linefeed combination for use on the prompt. Strictly speaking, the combination is unnecessary because VBA provides a constant for that combination (see the next section, VBA Constants). The example is there to show you that the function exists for combinations for which VBA does not provide. See the online Help for the *Chr* function to read about them in more detail.

VBA Constants. Constants are predefined data by VBA for use with functions and your procedures. They always have a character prefix of "vb". This routine contains the vbCrLf constant, which provides a carriage return–linefeed combination to a string. AutoCAD also uses constants that have a prefix of "ac". You can explore constants by using the Object Browser (the function key F2), selecting the AutoCAD Library in the upper-left pop-list, and typing **ac** in the pop-list below it. Then select the binocular button to display the search results. The bottom-right windowpane of the browser will display various constants as they apply to objects in AutoCAD.

Arrays. Arrays in VBA are similar to collections: they are designed to store multiple pieces of data. The difference is that their size must either be declared in advance, or you must keep adding on to them as the data grows. This is a slow process, and the main reason Microsoft recommends using collections in situations in which you want to store multiple pieces of data. While arrays do not support keys like collections do, they can allow multiple elements per row, while a collection can only have one element per row.

Autodesk uses arrays to hold points because points are always three single element rows and never change size. This means that when an object needs a point, it expects an array with the xyz point inside (one in each element of the array). There is a shortcut to creating the three single-element array each time. Instead of declaring an array variable, you declare a variant variable and let AutoCAD build the array in it for you when the operator selects a point.

 Note: To access the elements in an array, you must supply the subscript number for the element. Arrays are defined starting at zero by default (unless you actually declare it to start at a different number), and AutoCAD points must start at a zero subscript. To access the X element in a point array called Pnt, you would say Pnt(0), while you would access Y via Pnt(1), and Z via Pnt(2).

The GetPoint Method. The GetPoint method of the utility object returns a point unless the operator cancels when you request a point. Because the point variant will be empty when the operator cancels, it is always a wise move to check that there is a point in the variable.

IsEmpty Function. This function is used to determine when a variable has been initialized. It is ideal for determining when a point variable has been set. The AddLine exercise uses it to make sure the operator selected a starting point.

Object Update Method. Whenever you create a new object it will not appear on the screen until you use its Update method. Think of it as entity regeneration.

EDITING GEOMETRY

The steps to editing a piece of geometry are the same whether you want to edit a line, arc, circle, or any other entity type. You must first get the object and then put it into a container that matches the data type. You cannot put a line into a circle object, so it is important that you identify the entity type before you try to put it into a specific object data type. With that in mind, the steps to editing an entity are as follows.

EDITING AN ENTITY (WITHOUT FILTERING FOR THE OBJECT)

1. Get the entity into a generic object.
2. Check to see what type of entity it is.
3. Load it into the proper entity object container.
4. Make your changes to the entity object.

When using filters, whether the operator selects geometry, or you do, follow the next set of steps. You can skip the second step when you filter the selection of the geometry before it is selected.

EDITING AN ENTITY (WHEN FILTERING FOR THE OBJECT)

1. Get the entity using a filter for the entity you want, ensuring that only that type of entity is selected.

2. Load it into the proper entity object container.

3. Make your changes to the entity object.

The tricky part when you are first learning to deal with selected objects is to identify the entity while it is in a generic container. You must declare the correct object type before you try to fill it with the object you want to edit. Each entity object in the drawing has its own object type that you must use to work on it. There are several different techniques for working with the geometry objects.

One technique for acquiring an entity object is to have the operator select it using the GetEntity method of the utility object, as described earlier in the Operator Selection section. The problem with that technique is that you cannot filter out object types that you do not want to process, so you must check the object type after the operator selects it. Another similar technique is to use one of the Select methods and then filter out the object types you want by checking the ObjectName property of the generic object when it has an entity in it. When a SelectionSet is involved, you need to spin through it and process only the entity that meets your criteria (again by checking the ObjectName property). The technique that allows you the most control is the one that uses one of the Select methods with a filter attached, limiting the operator's ability to select only those entity types you specify.

The EditObject exercise is designed to illustrate the following concepts:

- Filters

- SelectionSets

- Editing geometry objects

Before you run this example place some lines, arcs, and circles in your drawing.

Code of Note in this Routine

Filters. This exercise demonstrates the use of a simple filter. It is possible to create very complex filters using the same techniques. For information on more complex filters, see the online Help index topic "Select method, example code."

The arguments for a filter must be in the format of two arrays. The first is an array of the integer type, which holds the DXF group code specifying the type of filter to use. The second array must be of the variant type and contain the value of the information to get.

The exercise sets the filter arrays to zero and circle, telling the filter to only select circles when the operator selects entities in the drawing. When used with any of the Select methods (the exercise uses the SelectOnScreen method), the filter ensures that your program will only get the entity type (a circle, in the exercise) you want in the returned SelectionSet.

Variant Data Type. This data type is a generic data type that will hold any other data type in it. It takes more space to store objects, so it is always best to declare the explicit data type when you know what it will be. Variants are often used as pointers to arrays.

SelectionSets. A SelectionSet is a special type of collection that Autodesk defined for your use. They work almost exactly the same as regular collections except their index starts at zero, whereas a regular collection's index starts at one. To use a SelectionSet you must first create it by adding it to the SelectionSets collection provided by Autodesk. Do you remember that collections can contain other collections? This is an example of that technique. All the tables in an AutoCAD drawing are collections containing other collections.

This exercise uses a toolbox routine to set up a named SelectionSet, because adding a SelectionSet name that is already there to the SelectionSets collection causes an error and will stop the program. The toolbox routine (called AddSelectionSet) traps any errors and handles them so that you do not have to worry about them. You just get your SelectionSet back ready to be filled.

SelectOnScreen Method. This is a method of the SelectionSet object that allows the operator to select geometry from the drawing (i.e. to load the SelectionSet with entities). It takes optional filter array information to limit the operator's ability to select only the geometry you specify. If the operator hits a cancel or does not select any geometry, the SelectionSet comes back empty (with a count of zero). For more information on the various Select methods, place your cursor on the method in the EditObject routine and hit the F1 key. The See Also link on the Help page takes you to a list of the other Select methods.

Editing the Geometry by Changing the Color of an Existing Circle. This exercise edits the geometry by simply changing its color. Once you have the entity in an object container, you can change any of the properties, use any of its methods, or just delete it.

TABLE ACCESS

Checking for an Existing Layer Definition

Layers are automatically added to the layer collection when you create them. There are times when you need to make sure that a layer already exists before you try to do

Note: As this exercise spins through the SelectionSet to process the selected object, it checks the ObjectName property of each entity to verify that it is a circle before it loads it into the circle object. This is unnecessary in a situation in which you use a filter, because the set will only contain the type you want. The code is there to illustrate how you would check if a filter had not been used with the SelectOnScreen method.

something with it. The exercise VerifyLayer illustrates the techniques for verifying that an item is already in a collection. The simplest way is to issue a call to the Layers collection using the Item method and then see if an error is thrown. An error means that the layer was not in the collection.

Code of Note in this Routine

The Item Method. You can access stored data inside a collection using its Item method. The method accepts a key parameter and returns the information associated with the key. When the key does not exist, an error is thrown. This is why it is important to have the On Error Resume Next statement active to trap the error by checking the state of the error object immediately after issuing the Item method to retrieve the data. If the error number is anything except zero, then an error occurred and the information was not loaded into the object.

The Error.Clear Method. Whenever an error is thrown and you trap it, you must clear the error out of the error object so it will be ready to catch the next error. The Clear method does the job for you. It does not need any parameters, so all you need to do is declare it.

Creating a New Layer

The Layers collection is one of the collections that represent the various tables in an AutoCAD drawing. All of them can be accessed directly from the ThisDrawing object. All of them have methods for adding new table items, deleting items, or retrieving items. The exercise CreateLayer is designed to illustrate how to create a new layer and set its color, and it introduces you to the VBA line-wrap feature.

The exercise creates a new layer, but when you run it again, nothing happens, despite the layer already existing. The Add method just ignores the problem and finishes without reporting any problem, because the layer is already there.

Code of Note in this Routine

Line Continuation. It can be extremely unwieldy to read or maintain code when it goes on much past the right-hand margin of your Code module window. VBA uses an underscore character (_) at the end of a line to wrap the line of code to the next line without causing an error in the interrupter. The string concentration of the message box's message shows how to wrap a line.

SYSTEM VARIABLES

The querying of system variables to make decisions and the changing of system variables are controlled by two methods of the ThisDrawing object.

Getting Them

The GetSysVar exercise illustrates how to get the Blipmode system variable from the drawing. You can retrieve any system variable by inserting it in place of the "blipmode" parameter.

 Note: Some variables are strings, not integers, so they would need to be placed in a string variable instead of in an integer variable.

Code of Note in this Routine

GetString Method. This method of the ThisDrawing object will retrieve any system variable defined by AutoCAD. For a complete list of the system variables, type **system variables** in the index of the online Help. An alphabetical list appears in the window.

Setting Them

The SetSysVar exercise illustrates how to set the Blipmode system variable from the drawing with a technique that will flip the variable back and forth each time you run the routine.

Code of Note in this Routine

SetString Method. This method of the ThisDrawing object will set any non-read-only system variable defined by AutoCAD.

Abs Function. The *Abs* function takes an integer of any value (positive or negative) and strips off the sign, revealing the number's magnitude. This comes in handy for flipping an integer variable used for True and False (1 and zero) from one setting to its other value, as shown in the following code:

```
Result = Abs(Blipmode + -1)
```

The *Blipmode* variable is either a 1 or a zero, depending on where it was turned On or Off. When you add a negative number to either of the settings, you get a zero or a −1. Using the absolute function on the resulting addition converts it to either a zero or a 1. The result is then pumped into the SetVariable method, and the system variable flips either from On to Off or from Off to On.

Test the setting of *Blipmode* by placing your cursor over it when the program is paused and then placing your cursor over the *Result* variable. You will observe that the old setting in the *Blipmode* variable is the opposite of the new setting in the *Result* variable.

PREFERENCES

Preferences in the object model consist of two distinct objects. The AcadDatabasePreferences object gives you access to all options from the Options dialog box that reside in the drawing. The AcadPreferences object gives you access to all the options from the Options dialog box that reside in the registry. The exercise called Preference illustrates several of the properties from each object.

 Note: Not only are two different objects used to access preferences, but they are accessed differently. The registry preference object is accessed from the Preferences property of the Application property of the ThisDrawing object, while the drawing preferences object is accessed from the Preferences property of the ThisDrawing object.

Code of Note in this Routine

Finding the preference option you want in the correct object is easy when you declare one of each object, as shown in the exercise, and then use the IntelliSense feature to see the list of available properties. Neither object has any methods. You can experiment in the Immediate window while the program is paused at the Stop function.

Getting the MenuFile Property. The MenuFile is a property of the AcadPreferences object's Files property (storing its data in the registry). It is sometimes handy to be able to see what menu file is currently loaded. Most of the options that appear on the Options dialog box are exposed in the AutoCAD object model. The options are organized just like they are in the Options dialog box folders.

Getting the LogFilePath Property. When you need to write out log files or other output, you can use the AutoCAD log file path. This property is also a property of the AcadPreferences object's Files property.

Getting and Setting a Global Lineweight. The global Lineweight property is a property of the AcadDatabasePreferences object (stored in each drawing). You can set any string variable to the property and do what you want with it, or you can set the property to a number corresponding to a legal lineweight setting (as shown in the exercise).

Using an Enumerated Constant. The AutoCAD 2002 version of VBA does not support creating your own enumeration constants, but AutoCAD has created many for your use with objects. Enumeration is an advanced concept that is essentially a list of integers that have each been given a corresponding name that can be used instead of the number. Because they are constants defined by AutoCAD, they all have the prefix "ac" as part of their name. Lineweight has a list of enumerated numbers, so you can see in English what you are assigning to the property. It is much easier to read and understand that acLnWtByLayer is setting the lineweight to

ByLayer than if you see a −1 for the setting. The exercise sets the global lineweight to acLnWt200, which means a lineweight of 200 (2.00 mm).

USERFORMS

UserForms (called forms in the Project window but UserForms in the Properties window) are special VBA objects designed for displaying control objects. A discussion on all aspects of forms can easily become a chapter in and of itself. This section should be considered a primer for creating forms.

Behavior

Forms in VBA are modal, meaning that you must hide them before you can work in the drawing and then you must show them again when you want to work on the form. Forms and their controls have properties and methods just like other objects in VBA, but they are special objects in that they also have events defined. An Event procedure is triggered by something happening when the form or one of its controls has focus (is selected). You place your code in the predefined Event procedure. The event codes most used are the click and dblclick (double-click) events. These are where you place your code to be processed when the operator selects a control. Other useful form or control events that can be coded and are common to most controls are as follows:

- Initialize code runs when the form is created. (Controls do not have an initialize event.)

- KeyDown code runs when an operator presses a key while a control has focus.

- KeyUp code runs when an operator releases a key while a control has focus.

- KeyPress code runs when an operator presses a key while a control has focus. It is used to test keystrokes for validity immediately or to format characters as they are typed in a TextBox or ComboBox.

- MouseUp code runs when an operator releases a mouse button (which you can determine).

- MouseDown code runs when an operator presses a mouse button down (which you can determine).

- Terminate code runs when all references to the form are removed from memory. (Controls do not have a terminate event.)

When you work in the Code module window on a control's Event procedure code, you will find that the Object List pop-list contains all the controls defined on the form (as well as the form itself and the General section). The Procedure List contains the predefined Event procedures for each selected control, as well as any new ones you add.

One control's event code can initiate another control's behavior or properties. This type of behavior is explored in more detail in the Control Interaction section, and the Forms exercise provided for this section.

Controls

A control object can only exist on a form object. Controls are where the operator interfaces with your code. Most have a click event for placing code in response to the operator's selection. Some have a default change event in place of the click event, and a few only have the change event with no click event. Whatever the default is, you can place your code in whichever event code procedure you want. Some of the most used and useful controls are as follows:

- The Label control displays information. This control is also good for storing data in its Tag property for use by other controls or procedures in your program.

- The TextBox control is used to edit ASCII text data. This control allows the operator to enter in text strings for you to process.

- The ListBox control displays a list of items and allows the operator to select one or more.

- The ComboBox control combines the features of the TextBox and ListBox controls.

- The CheckBox control allows the operator to select between a state of On or Off (True or False).

- The Frame control groups similar controls for display purposes. This control can also be used to store the choice made by the controls inside it for later processing.

- The OptionButton control is used with at least one other OptionButton control in a frame to give the operator a choice of items. Only one button can be selected at a time when these buttons are grouped together in a frame.

- The CommandButton control is used to run code when the operator selects the button. A typical use of this control is to exit the program.

There are several methods for finding out more detail about the controls and their properties. One is to select a control on the form and hit the F1 key. This will open the online Help for the control, where you can read about the control, its properties, and it methods using the links to those topics at the top. Another method of getting help on a specific property is to first select the control on the form and then select a property in the Property window. Once the property item is selected use the F1 key, you can go straight to that property's online Help.

To start coding a control is a simple matter. Drag the control onto the form from the Toolbox Controls folder and double-click on it. The editor will automatically switch you over to the default event procedure for that control so you can immediately start placing code. Once the procedure code is exposed in the Code module window, you can use the Object list and Procedure list pop-lists to place other predefined procedures or create your own.

Because the controls on forms are not sequential, the operator can move around among them as they please. Building forms can be as much an art as a science. Besides understanding how the controls work independently and together, you also need to understand how the operators want to use them. The placement of the controls on a form should not look awkward, nor should the controls be grouped in a random fashion. Organize the controls on the form in a logical fashion for the operator's use.

Control Interaction

When you design your form, you need to keep in mind the information you need from the operator to do your job and how the operator will interact with the controls on the form. You do not want the operator to be able to proceed until all the necessary information is gathered. To this end you need to design the controls to only be available (enabled) once you have the necessary information from another control. This keeps the random nature of the form somewhat intact and also ensures that the information you need to process or continue working with is available. You should never have the OK button enabled unless you know that all the information you need to process is present on the form.

The Forms exercise illustrating how to program the controls is simple to use, but you need to get the editor in the correct state before you can proceed. Simply start the form by double-clicking on the UserForm1 in the Forms folder in the Project window. The Visual Basic Editor should look something like Figure 10–8 before you are ready to run the program.

The Forms exercise is designed to illustrate the following concepts:

- Interaction between controls
- Getting input from the operator at the command line and placing it in a text box
- Setting the properties of controls
- Exploring all the controls listed in the Control section

Once the form is displayed in the Code module, select the Run Sub/UserForm button from the Standard toolbar. The form appears in front of the AutoCAD drawing. You can now select different controls and step through the code to see how it works and how some controls interact with other controls. You can experiment in any order you wish. Each event has a Stop statement in it so you can look around as you explore.

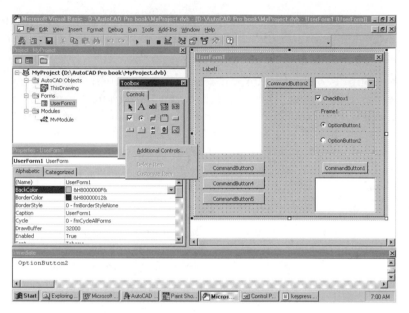

Figure 10–8 *The VBAIDE environment with a form open.*

 Note: When a form is in the Code module for development, the Toolbox window should appear (as shown in Figure 10–8). If it does not, you can expose it by selecting the Toolbox menu item under the View pull-down menu. Figure 10–8 also shows the Additional Controls menu that pops up when you right-click on the toolbox. This menu allows you to add other controls to your toolbox. The form shown in the figure has the actual default control names and captions that were assigned when they were created. It is up to you to change their names in the Properties window or through code as is done in the Set_Defaults procedure of this exercise. No two controls can have the same name.

 Note: When using the technique of stepping through the code by hitting the F8 key for each line, you will need to select the Run Sub/UserForm button again when you get to the End Sub statement to get back to the form displayed in AutoCAD. Another method is to continue past the End Sub statement and then use the Alt+Tab key combination to shift back over to the form running in AutoCAD.

Code of Note in this Routine

UserForm_Initialize Event. This event code is triggered after a form object is loaded but before it is shown. This event is used to load the controls of the form before it is shown. The form object in this exercise calls a separate procedure (called Set_Defaults) to do the loading.

Set_Defaults Procedure. This procedure is used to load the form before it is shown for the first time. This procedure illustrates how to load a control's properties (captions and default values) from code.

CommandButton1_Click (Get a String Button) Event. This event code uses the GetString method (examined in detail later in this section) to get input from the operator at the command line and place it into the TextBox control (the lower-right control on the form).

CommandButton2_Click (>>> Button) Event. This event code takes all selected items in the ListBox and copies them over to the ComboBox. When you get back to the dialog box, notice that the TextBox in the lower-right corner of the form reflects the first item in the ComboBox. This is because the first item in the ComboBox is placed in view by the code in this event, just as if the operator from the ComboBox had selected it. That means that the Combobox_Changed event was fired off and its code ran, putting the information into the TextBox (and the TextBox's Change method then fired and executed any code that was there to execute).

This is an example of how manipulating other controls can be tricky. You need to be aware that one event can trigger another. It is even possible to get into an infinite loop this way, where you cannot break out or do anything.

CommandButton3_Click (Load ListBox Button) Event. This event code uses the Clear method (explained later in this section) to clear out the ListBox control and then calls the Load_List procedure (also explained later in this section) to reload the list with data.

CommandButton4_Click (Clear Selections Button) Event. This event code spins through the ListBox items and clears any that are selected. It uses the ListCount and ListSelected properties explained later in this section.

CheckBox1_Click (Use ComboBox Checkbox) Event. This event code enables or disables the ComboBox control depending upon whether or not it is checked. It uses the Value property explained later in this section to check its own status, and sets the Enable property (also explained later) of the ComboBox control based upon its findings.

CommandButton5_Click (Clear Listbox Button) Event. This event code clears the ListBox using its Clear method (explained later in this section).

ComboBox1_Change Event. This event code takes the Value property (explained later in this section) of the ComboBox and places the selected item's text into the TextBox.

OptionButton1_Click (Left option Button) Event. This event code places the information into the Frame control's Tag property and into the TextBox's Text property. Both properties are explained in detail later in this section.

OptionButton2_Click (Right Option Button) Event. This event code places the information into the Frame control's Tag property and into the TextBox's Text property. Both properties are explained in detail later in this section.

The Frame Control. The Frame control is designed to group option buttons together. Option buttons are grouped together by design because one by itself is not much use. Another use of the Frame control is to use its Tag property to store the currently selected option button in the frame. The event code for both OptionButton1 and OptionButton2 illustrates the data-storing technique. The Set_Default procedure illustrates one technique for using the data. Anytime you want to know the setting of the option buttons, you only have to query the frame's Tag property.

Load_Listbox Procedure. This procedure illustrates the AddItem method used to load a ListBox control. The AddItem method is explained in more detail later in this section.

Value Property. This property is used by controls to store a value. Both the Check-Box and ComboBox use this property to store selected information.

Text Property. This is another property used by controls to store a value. The TextBox uses this property to store the information typed into it.

The With Statement. This shortcut executes a series of statements related to the object. It takes an object as the parameter and exposes the object's methods and properties within the block of code before its End statement. It is a shortcut because it saves typing time. Instead of having to type the object with a period to expose the methods and properties, you only have to type the period and then the IntelliSense feature shows you the possible choices. The Set_Defaults procedure illustrates how this statement works.

The Me Keyword. Whenever you are in the Form module coding procedures, you can use the Me keyword to reference the form. It is a shortcut, so you do not have to type the actual name of the form each time you want to reference it. In most situations you do not even need to reference the form when you access one of its controls properties. VBA will figure out that the control is on the form by itself. It is simply a good coding habit to include the explicit reference so any other programmers viewing the code will understand that the control is part of the form.

The Hide and Show Methods of a Form. Because VBA forms are modal, you cannot have them showing when you want to get data from the drawing (any kind of data input from the operator using any kind of data acquisition method). The Hide method of the form temporarily puts the form to sleep (i.e., minimizes it). Once the operator has supplied the desired information, the Show method is used to bring the form back to life (i.e., restore it to the screen).

Clear Method. Often it is necessary to clear out the information in a ListBox or ComboBox control before you reload it. Both controls have this method to perform the task.

Selected Property. The ListBox control has this property for determining if a given item in the list is selected. The Clear ListBox button's event code uses this property to clear all selections from the list.

AddItem Method. This method adds one line of data to the item list for both the ListBox and ComboBox controls. Both the >>> button event code and the Load_List procedure use this method to load data into the controls. This method takes two optional parameters. The first parameter is the data to place in the control. The second parameter specifies the position in the list at which the new item will be placed. When you use this method without any parameters, a blank line is added to the control list.

Note: The first parameter is optional. When you use this method without any parameters, a blank line is added to the control list.

Enabled Property. All controls have this property. It tells the form whether or not to allow the operator access to the control. Setting the control's property to True allows operators access, and setting it to False keeps them from accessing it. The "Use ComboBox" CheckBox event code uses this property to set the ComboBox's Enable property. The >>> labeled button event code uses this property to verify whether or not the ComboBox's Enable property is set.

SetFocus Method. The >>> button event code uses this method to set the focus on the ComboBox control and then back to the >>> button. Anytime an operator selects a control the control takes the window's focus. When programming a control it is sometimes necessary to have the focus set to the control in order to manipulate it. This method takes no parameters.

ListCount Property. This property applies to both the ListBox and ComboBox controls. It is used to find out how many items are loaded into the control. Both the >>>> button event code and the Clear Selections button event code use this property to control spinning through their list of items.

ListIndex Property. This property identifies the currently selected item in both the ListBox and ComboBox controls. You can set a selected item using this property by supplying an index number corresponding to the item in the list. You can also query the property to get the current selected item's index number. A negative number is returned when nothing is selected in the list.

Tag Property. This property is present on most controls. It is an additional place to store data on the control object. Unlike the Text and Value properties, it is not used by any of the controls as a default or standard storage place, so you do not have to worry about your data messing something else up. Both the OptionButton event code and the Set_Defaults procedure use this property.

Caption Property. Many controls have this property. It displays control information to the operator on the form. The Set_Defaults procedure uses this property to set all those controls on the form that have the property.

GetString Method. The utility object has this method for gathering data from the operator on the command line. This method takes two parameters. The first is a mode telling the method whether or not to allow spaces in the operator's response. (A one tells it to accept spaces and a zero to not accept spaces.) The second parameter is the prompt that the method will display on the command line.

Selecting an Item in a ListBox or ComboBox Control. When the operator selects an item from one of these controls, the Value property is set to the item selected, and the change and click events for that control are triggered unless the MultiSelect property for that control is NOT set to MultiSelectSingle. When The MultiSelect property is set to MultiSelectMulti or MultiSelectExtended, the Value property is not set, and the Change and Click events are not triggered upon an item or items being selected in the control.

ComboBox1_Change Event. The change event code gets triggered when an item is selected from a control that has its MultiSelect property set to MultiSelectSingle. You must be careful coding a change event when you want to do something to another control that will trigger its change event code. You can get into an infinite loop where the two controls keep triggering each other. It can be very hard to break out of these types of loops.

SUMMARY

In this chapter you were exposed to the basics of programming in VBA, the development environment for creating programs, the AutoCAD object model, debugging your program techniques, and techniques for testing your code as it runs. It is hard to adequately cover so complex a subject as VBA in just one chapter, but you should now be at least comfortable with the concepts, the working environment, and the complexity of the language.

You can use the techniques covered in this chapter to walk though code, to examine it and see what happens as it runs. the techniques illustrated in this chapter can be usedto examine the sample programs supplied with AutoCAD by Autodesk. The following are two excellent projects for reviewing and learning more about how VBA works inside AutoCAD:

- acad_cg.dvb
- example_code.dvb

You can also explore any of the other sample projects in the Sample\VBA folder in the AutoCAD directory.

SECTION

V

Managing
AutoCAD

CHAPTER 11

Determining ROI—Return on Investment

One of the biggest challenges CAD managers face is effectively communicating CAD needs to upper management. The CAD manager is typically well versed in CAD and intimately familiar with users' needs for new hardware and software and for training. The CAD manager views fulfilling users' needs as critically essential to the company's welfare, and typically approaches fulfilling CAD-related needs with a sense of urgency. From the CAD manager's perspective, CAD-related needs are *very* important. In contrast, upper management is usually unfamiliar with CAD and therefore typically does not perceive fulfilling users' CAD-related needs as having the same level of importance. However, what upper management *does* perceive is this: CAD-related needs are never-ending and very expensive. From upper management's perspective, CAD-related needs must be controlled and limited to reasonable levels of recurring costs. These two perspectives often cause conflict between the CAD manager and upper management, with the CAD users—and ultimately the company—suffering the consequences of outdated hardware and software, and little-to-no training.

Of course, both perspectives are accurate. The truth is that meeting CAD-related needs *is* essential to a company's survival, and CAD-related needs *are* never-ending and can be very expensive. Therefore, it is imperative that CAD managers *effectively* express CAD-related needs to upper management. To do so requires not only understanding the challenges faced by both CAD users and upper management; it also requires that CAD managers express CAD-related needs in a language that upper management understands. An excellent way to express needs to upper management is through the language of finances. In other words: Money talks.

In this chapter you will learn how to use the language of finances by calculating the Return on Investment (ROI) of your CAD-related needs. You will also be introduced to Capital Budgeting Analysis, which helps you understand the time value of money, a concept that often drives the financial decisions that upper management makes. Through ROI and Capital Budgeting Analysis, you not only effectively communicate the benefit of new technology and training to upper management in a language that they recognize, you demonstrate that you understand and appreciate the challenge upper management faces: keeping your company profitable and competitive.

THE BOTTOM LINE

In my experience, upper management is neither a friend nor a foe of CAD technology. Upper management simply understands that CAD is a necessary part of their industry. What upper management faces, however, is balancing the needs of *all* aspects of their business, not just CAD-related needs, and it must do so within a limited budget. When faced with a request that costs the company money, upper management must determine whether the requested funds, or *expenditures*, are financially prudent. Therefore, a requested fund for anything is commonly viewed as a financial investment. Simply stated, upper management must determine what the pay-off is for spending money. If an expenditure makes money (increases profits), then the company will strive to fund that expenditure. However, if an expenditure costs money, then the company must give serious thought as to whether the other, less-tangible benefits of the expenditure justify its funding.

Some expenditures are no-brainers, such as telephones. Communication between clients, vendors, and employees is essential, and telephones are an excellent, simple-to-use communication tool. In regard to telephones, calculating ROI is not that essential an exercise, because upper management knows that telephones are absolutely necessary for their company's survival and must be acquired. On the other end of the spectrum are expenditures that are not necessary for the company's survival, such as company-sponsored picnics or holiday parties. These events cost the company money, and their value is represented by the morale boost this type of event produces, which may increase the bottom line in hard-to-measure, less-tangible ways. That is, happy employees are likely to be more productive than unhappy employees.

CAD-related expenditures, however, lie somewhere in between telephones and parties. While it certainly can be argued that design firms must have CAD systems in order to function, just as they must have telephones, questions arise as to how much to spend on hardware and software, as well as on training. For example, does your company need an expensive specialized application such as Land Desktop or Mechanical Desktop, or can it get by with a simple and much less expensive CAD application such as AutoCAD LT? And do CAD users really need training, given that training is typically expensive?

It is important to understand that when upper management requests justification for CAD-related expenditures, it is not because they are uncaring or indifferent to CAD needs. Upper management is responsible for ensuring that the company is healthy and profitable. Therefore, upper management simply needs to see that the expenditures for CAD-related needs increase the company's profit margins and are therefore justified. When you show that a CAD-related expenditure will make the company more profitable, then you make it easier for upper management to approve your request.

In ROI you have a tool that communicates CAD-related needs to upper management. By calculating an ROI, you clearly and concisely show upper management why, for example, a new large-format plotter is justified, or why sending staff to a one-week CAD Training Camp is prudent. By using ROI as a communication tool, you demonstrate to upper management that by meeting CAD users' needs, they increase the company's bottom line.

JUSTIFYING EXPENDITURES

Every time a new version of AutoCAD is released, CAD managers face the same predicament: whether or not to upgrade to the newest release. For some, there is no question at all—their company intends to stay on the leading edge of technology and will therefore upgrade as soon as the new release is available. However, for most, the question of upgrading is not nearly as easy when the key concern is determining whether the expenditure will pay off, ultimately increasing the company's bottom line. So for most companies, deciding whether or not to upgrade is viewed as an investment where the decision to upgrade is directly related to the financial benefits the company expects to receive from the upgrade and to how quickly those benefits are realized.

One of the best ways to demonstrate the value of an investment in CAD is to calculate its ROI. The ROI provides your company's upper management with objective, quantifiable information that helps them decide whether upgrading their CAD systems is financially prudent. By calculating ROI, you show the financial benefits of upgrading to a new release, and the increased profits your company can anticipate.

Calculating an expenditure's ROI is accomplished using the following formula:

$$ROI = \{\{A - [B / (1 + E)]\} \times (12 - C)\} / [A + (B \times C \times D)]$$

where

- **Variable *A* equals the total system cost (in dollars).** This represents the estimated cost of software and can include the cost of hardware, peripherals, and other services you will purchase to equip a single CAD seat. You should also include the cost of training new users how to use the software. If you are considering purchases at different cost levels (such as a Compaq system versus an HP), you should make a separate ROI calculation for each scenario.

- **Variable *B* equals the total monthly labor cost (in dollars).** This represents the estimated total monthly cost of a typical employee currently performing design and drafting work. You should include salary and benefits but not overhead.

- **Variable *C* equals the required training time (in months).** This represents the estimated amount of time required for the employee to master the new software. (This time is also called "learning curve"). Although this value varies from one employee to another depending on skill level, choose an average value that represents the time required for users to become efficient.

- **Variable *D* equals the training productivity loss (in percent).** This represents the anticipated loss of productivity during the training time. In other words, how much less productive is the employee while learning the new version of AutoCAD? Will the employee be 50% less productive? 30%? 70%?

- **Variable *E* equals the final productivity gain (in percent).** This represents the productivity increase that you expect to realize after training is completed. In other words, how much more productive will this employee be after learning the new version of AutoCAD?

As you might suspect, some variables are easy to determine, such as variables *A* and *B*. Others are more subjective, such as variables *D* and *E*. While determining subjective values is not easy, you should rely on past experience to make reasonable estimates.

AN ROI CASE STUDY

To demonstrate how to calculate ROI, let us assume the hypothetical case of a firm buying a complete CAD system, which includes hardware and software. In this hypothetical, a new employee is hired to do full-time drafting at a monthly labor cost of $3,300. After buying the CAD system for $10,000, the firm lets the employee spend three months training on the system. During this training period, the employee is totally non-billable, spending his entire time learning the new CAD system.

After the three-month training period, the employee can use CAD efficiently. Because of his new CAD skills, the employee increases his overall productivity by 25% from when he was originally hired. Given these values, let us calculate the ROI:

- Variable *A* (total system cost in dollars) = $10,000

- Variable *B* (total monthly labor cost in dollars) = $3,300

- Variable *C* (required training time in months) = 3

- Variable *D* (training productivity loss in percent) = 100% (three months learning the system)

- Variable *E* (final productivity gain in percent) = 25%

Therefore

$$ROI = \{\{\$3,300 - [\$3,300 / (1 + 0.25)]\} \times (12 - 3)\} / [\$10,000 + (\$3,300 \times 3 \times 1.00)]$$

which means

$$ROI = 0.298 = 30\%$$

Therefore, this firm would anticipate a 30% return on investment during the first year.

Note: The percentage values are inserted into the formula in their decimal equivalents. Therefore, 100% is entered as 1.00, and 25% is entered as 0.25.

What ROI Means

In the previous example we calculated a 30% ROI. So what does a 30% ROI mean? It means that if the firm spends $10,000, in one year they well get back 30% of that expenditure, which equals $3,000. Given this ROI, do you think it is financially prudent for the firm to spend $10,000 on hardware, software, and training? To help you answer that question, ask yourself if you are willing to spend $10,000 to get back only $3,000. I know I am not. So does this mean that the firm would reject the request for funds? Not necessarily.

In the above case study, the estimated 30% ROI is for Year 1 *only*. What about the second year that the employee continues working? And what about the third, fourth, and fifth years of employment? Each year, the firm continues to benefit from the employee's 25% productivity increase (variable *E*). More importantly, the other costs (such as those for hardware, software, and training), as well as the learning curve, are static, occurring only in Year 1. To understand the effects of ROI on the firm's original $10,000 investment in subsequent years, we must calculate each year's ROI. But before we can calculate ROI for subsequent years, we need to understand how the ROI formula calculates ROI.

Understanding the ROI Formula. To apply the ROI formula to investments beyond Year 1, we need to modify the formula slightly, which means we need to understand what the formula is calculating. Fortunately, for our purposes, we can keep this explanation very simple.

From our previous discussion, we know that the ROI formula is

$$ROI = \{\{B - [B / (1 + E)]\} \times (12 - C)\} / [A + (B \times C \times D)]$$

which we can separate into two main parts. The first part appears to the left of the second division (/) sign in the equation's numerator as

$$\{\{B - [B / (1 + E)]\} \times (12 - C)\}$$

and represents our expected *productivity gain* in dollars. The second part appears to the right of the second division sign in the equation's denominator as

$$[A + (B \times C \times D)]$$

and represents our expected *total cost*. Both the numerator's productivity gain and the denominator's total cost are cumulative, which means that for the total number of years that you are calculating ROI, you add all productivity gains together and add all costs together. So, to calculate the ROI for Year 2, you add the productivity gains for Years 1 and 2 together, then add the total costs for Years 1 and 2 together, and then divide the productivity gains by the total costs. To demonstrate, let us first calculate all productivity gains:

Year 1 productivity gains = $\{\{\$3,300 - [\$3,300 / (1 + 0.25)]\} \times (12 - 3)\} = \$5,940$

Year 2 productivity gains = $\{\{\$3,300 - [\$3,300 / (1 + 0.25)]\} \times (12-0)\} = \$7,920$

Therefore, the cumulative productivity gains = \$5,940 + \$7,920 = \$13,860. Next, let us calculate all total costs:

Year 1 total costs = $[\$10,000 + (\$3,300 \times 3 \times 1.00)] = \$19,900$

Year 2 total costs = $[0 + (\$3,300 \times 0 \times 0.00)] = \$0$

Therefore, the cumulative total costs = \$19,900 + \$0 = \$19,900. Finally, we divide the productivity gains by the total costs:

$$ROI = \$13,860 / \$19,900$$

which means

$$ROI = 0.696 = 70\%$$

So, in Year 2, the firm's cumulative return is 70% of its initial \$10,000 investment, which equals \$7,000. By continuing in this fashion, calculating each year's cumulative ROI, you will identify the year in which the firm receives back its original investment. In this example, it occurs in Year 3 when the ROI reaches 109%. Once a firm achieves 100% ROI, it means that it has recouped its original investment, and any subsequent years are pure profit.

 Note: The recouped investment dollars are known as cash flow. In the above example, by the end of Year 2, the firm will receive a cumulative total of $7,000 cash flow: $3,000 cash flow for Year 1 and $4,000 cash flow for Year 2.

So far we have reviewed ROI from the perspective of purchasing an entire CAD system, both hardware and software, as well as for significant training time and learning curve. This is a rather harsh set of values compared to the costs expected for upgrading existing users to a new version of AutoCAD, as discussed in the following section.

 Tip: To calculate the average annual ROI, divide cumulative ROI by the number of years. For example, to calculate the ROI for Year 3, divide 109% by 3, which equals 36% annual ROI.

UPGRADE ROI

The above case study uses numbers that represent purchasing an entire CAD system, and then training an employee who is not billable for three months. In contrast, when you upgrade existing users to a new AutoCAD release, both the total system cost and the productivity lost to training (the learning curve) are typically much smaller than those shown in the previous example. Because these ROI variables are much smaller, the first-year ROI can be very impressive. For example, let us assume the following values for calculating the upgrade ROI:

- Variable *A* (total system cost in dollars) = $800 (upgrade cost to new AutoCAD version)
- Variable *B* (total monthly labor cost in dollars) = $3,300
- Variable *C* (required training time in months) = 2
- Variable *D* (training productivity loss in percent) = 20%
- Variable *E* (final productivity gain in percent) = 10%

Therefore

$$ROI = \{\{\$3{,}300 - [\$3{,}300 / (1 + 0.10)]\} \times (12 - 2)) / [\$800 + (\$3{,}300 \times 2 \times 0.20)]$$

which means

$$ROI = 1.415 = 142\%$$

Therefore, this firm would anticipate a 142% return on investment during the first year of the upgrade, assuming a training time (learning curve) of two months, a 20% productivity loss during the training time, and only a 10% productivity gain. This means that the firm would recoup its initial investment within nine months and gain more than 40% return on its investment dollars in the first year. So, to financially justify upgrading to a new release of AutoCAD, even with a learning curve as long as two months, all you need is a relatively small productivity gain, such as 10%. As you can see, calculating ROI can quite easily result in rates of return that justify upgrading to a new release of AutoCAD.

 Note: Autodesk provides an excellent tool for calculating ROI. Created by Autodesk's Bob Ng, the ROI2000.xls Excel spreadsheet, which is included on the accompanying CD, includes a detailed explanation of all of its variables and how to use the spreadsheet.

AN INTRODUCTION TO CAPITAL BUDGETING ANALYSIS

While you have learned how to calculate ROI and how to generate numbers that can prove that upgrading to a new version of AutoCAD is financially viable, the ROI rate of return does not tell the whole story, especially when the company has limited funds and different departments vying for those limited funds. Even though you present ROI numbers that justify your proposed expenditures, other departments may produce equally justifiable ROI numbers in their proposals. This presents upper management with a dilemma: how to decide which proposals to fund when faced with several choices and a limited budget. This is where capital budgeting analysis comes into play.

Capital budgeting analysis provides managers with a tool that helps them identify which proposals to fund. Of the proposals presented to managers that justify funding, capital budgeting analysis lets managers rank the proposals in terms of overall profitability and cash flow potential. Through capital budgeting analysis, managers can assign a ranking to proposals and thereby make decisions that will maximize corporate profits and cash flow.

There are three common methods for analyzing financial investments:

1. **Payback Method:** The number of years required to earn back the original investment

2. **Net Present Value (NPV)**: The present value of all related cash flows discounted at the cost of capital

3. **Internal Rate of Return (IRR)**: The interest rate at which the net present value of all related cash flows is equal to zero

In the following sections these three methods for analyzing financial investments are compared against using a simple case study.

A CAPITAL BUDGETING ANALYSIS CASE STUDY

Assume that your firm is reviewing three proposals for improving three different operations within the company, and that each proposal requires a single initial investment of $1,000. Also assume that the projected cash flows (calculated from ROI) generated by the three proposals are as shown in Table 11–1.

Table 11–1 shows the anticipated cash flows each year following the initial $1,000 expenditure. For example, in Year 1, Proposal A generates $500, while Proposal B generates $100, and Proposal C generates no cash flow at all. For each year after the investment, the anticipated cash flows are calculated (using the ROI formula presented earlier in this chapter), and then totaled after five years. Table 11–1 shows that after five years, Proposal A generates $1,300 total return, while Proposal B generates $1,500, and Proposal C generates $1,700. By subtracting the initial expenditure of $1,000 for each proposal from the total returned, the net cash flow is calculated, with Proposal A generating $300, Proposal B generating $500, and Proposal C generating $700 total net cash flow.

Given the values in this case study, in which proposal do you invest: Proposal A, totaling $300 net cash flow; Proposal B, totaling $500 net cash flow; or Proposal C, totaling $700 net cash flow? As noted earlier, there are three common methods for analyzing financial investments. Let us look at the first, the Payback Method.

Using the Payback Method

The Payback Method determines the period (or number of years) it takes a firm to recover its initial investment. The payback period represents the year in which the $1,000 funding is recouped, as shown in Table 11–2. In the case study here, the payback period occurs in Year 3 for Proposal A, in Year 4 for Proposal B, and in Year 5 for Proposal C. This means that Proposal A pays back the firm's $1,000 expenditure in the shortest time frame, while Proposal C takes the longest.

Table 11–1: *Net Cash Flow Projections*

Year	Proposal A	Proposal B	Proposal C
0	-$1,000	-$1,000	-$1,000
1	500	100	0
2	400	200	0
3	300	300	0
4	100	400	0
5	0	500	1,700
Total net cash flow	$300	$500	$700

Table 11–2: *Payback Period*

Year	Proposal A	Proposal B	Proposal C
0	-$1,000	-$1,000	-$1,000
1	-$500	-$900	-$1,000
2	-$100	-$700	-$1,000
3	$200[a]	-$400	-$1,000
4	$300	0[a]	-$1,000
5	0	$500	$700[a]

a. Year in which payback occurs

Table 11–2 shows the net cash flow projection for each proposal for each year. For example, in Year 1 Proposal B projects a net cash flow of -$900, which represents the difference between the cash flows generated in Year 1 for Proposal B ($100 as shown in Table 11–1) and the initial $1,000 expenditure.

Some firms use a payback period to determine whether an expenditure is warranted. For example, if a firm establishes a three-year payback period as a requirement for funding, then it would approve only Proposal A and reject Proposals B and C because only Proposal A lets the firm recoup its expenditure within three years.

Although the Payback Method is easy to calculate, it does not necessarily lead to the best financial decision. As the previous example shows, it ignores cash flows beyond the payback period. The value of proposals that generate long-term cash flows with sizable returns beyond the payback period are completely unrecognized by the payback method, which can be a serious shortcoming. As shown in Tables 11-1 and 11-2, while Proposals B and C take longer to pay back the firm's initial expenditure, they both result in significantly greater net cash flows.

Another shortcoming of the Payback Method is that it ignores the firm's cost of capital. The cost of capital applies to both the cost of borrowing money to make the investments as well as to the opportunity to earn interest on savings or profits generated by the proposal. This is why many firms review proposals by calculating their Net Present Value (NPV) and their Internal Rate of Return (IRR). However, before the NPV or IRR can be calculated, the time value of money must first be determined, as discussed in the following section.

The Time Value of Money and Discounted Cash Flow

As shown in the previous section, while it is easy to calculate, the Payback Method does not consider the overall cash flows generated by an expenditure, ignoring the total time period during which an expenditure generates cash flow. Also, the Payback Method ignores the cost of capital—which represents the price the firm pays for borrowing money for an expenditure—as well as the amount of interest a firm can earn on cash flows generated by an expenditure.

Because of the shortcomings of the Payback Method, financial analysts developed a variety of *discounted cash flow* techniques that help them recognize the *time value of money*. Through these techniques, analysts can account for the overall time period during which an expenditure generates cash flows, and they can compare the value of cash flows generated in the future to the value of funds today. In other words, these techniques let us determine if $1 received today is preferable to $1 received a year from today. For the case study discussed here, these techniques can help us better determine which proposal is preferable: Proposal A, which provides a three-year payback, or Proposals B and C, which produce greater cash flows.

The Value of Today's Dollar. To help us determine which proposal represents the most profitable expenditure, we must first understand the current and future value of a borrowed dollar. Let us assume that we can borrow money from a bank at an annual interest rate of 10%. Given this information, we can easily calculate the future value of today's dollar.

Suppose that we decide to borrow $1 from the bank for one year. This means that at the end of one year, we must repay the bank $1 plus 10% interest on the borrowed $1, which means we must repay a total of $1.10. (This is calculated by multiplying $1 by 1.10, which equals $1.10.)

Tip: To calculate the interest rate multiplication factor, divide the interest rate by 100 and then add 1. So, if the interest rate is 10%, then (10/100) + 1 = 1.10. Therefore, a 10% interest rate has an interest rate multiplication factor of 1.10.

Suppose, instead, that we decide to borrow $1 from the bank for two years. This means that at the end of two years, we must repay the bank $1 plus 10% interest for two years on the borrowed $1, which means we must repay a total of $1.21. (This is calculated by multiplying $1 by 1.10, which equals $1.10 at the end of Year 1, and then multiplying $1.10 by 1.10, which equals $1.21 at the end of Year 2.)

Note: Interest rates are applied annually. So, when you borrow money at an interest rate of 10%, if you borrow the money for two years, you will pay 10% for the first year, and then pay another 10% for the second year.

Continuing in this fashion we can calculate the amount due at the end of a loan period for any given length of time and thereby develop a schedule of repayments, as shown in Table 11–3.

The values in Table 11–3 are *equalized* to account for the time value of money. In other words, the table indicates that $1 received today is equal in value to $1.10 received one year from today, which is equal to $1.61 received five years from today, which is equal to $2.59 received ten years from today. These values represent how much a future dollar is worth relative to today, assuming we can borrow funds at a 10% interest rate.

 Note: The discount factors shown in Table 11–3 are based on a 10% interest rate and vary based on the interest rate used.

By equalizing the time value of money, we can compare the future cash flow earnings of the various proposals relative to the value of today's dollar. In financial terms, we are *indifferent* to the various yearly cash flows of these proposals. Each proposal's yearly cash flow represents equal value when the amount of cash generated and the future date when it is received are compared. By equalizing the values of future dollars, we create a level playing field in which to compare the true earning potential of each proposal's total cash flows.

Table 11–3: *Repayment Schedule for a $1 Loan at 10% Annual Interest*

Loan is Due End of	Total Amount Due
Year 1	$1.10
Year 2	$1.21
Year 3	$1.33
Year 4	$1.46
Year 5	$1.61
Year 6	$1.77
Year 7	$1.95
Year 8	$2.14
Year 9	$2.36
Year 10	$2.59

Note: The total amount due at the end of the loan period is commonly referred to as the "discount factor" in financial terminology.

Tip: You can exponentially calculate the amount due at the end of a given loan period by raising the loan's interest rate multiplication factor to the power of the loan's period. For example, when you raise the interest rate multiplication factor (1.10) to the power of 8 (for an 8-year loan period), the result equals 2.143.

Applying the Discount Factor . Table 11–3 shows the time value of money, where the value of future dollars (future cash flows) is equalized to account for the loan period. Given these equalized values, or discount factors, we can view the value of future cash flows relative to today's dollar value. For example, referring to Table 11–1, to view the value of cash flows generated by Proposal A in Year 4 ($100) relative to today's dollars, you divide $100 by the discount factor for Year 4 shown in Table 11–3. Therefore, divide $100 by $1.46, which equals 68.493. This means that the $100 cash flow generated in Year 4 is approximately equal to $68 in today's dollars.

Another way financial analysts apply the discount factor is by *multiplying* the cash flow value by the inverse of the discount factor. The inverse of the discount factor simply means you divide 1 by the discount factor. For example, using the discount factor values from the previous example, dividing 1 by $1.46 equals 0.68493, which you would then multiply by the $100 cash flow in Proposal A's Year 4 to get approximately $68. Multiplying by the inverse of the discount factor is an easy way to calculate the value of future cash flows relative to today's dollar value.

To understand better the meaning of the inverse of the discount factor, think of it in terms of a loan whose final payment is equal to $1. So, rather than borrowing $1 and then calculating the total amount you must repay at the end of the loan period, you work backwards and estimate the amount you must initially borrow in order for the payoff at the end of the loan period to equal $1. Again, these values are easily derived by calculating the inverse of the discount factor, as shown in Table 11–4.

In Table 11–4 the values in the Inverse Discount Factor column represent the amounts you must initially borrow in order for the payoff at the end of the loan period to equal $1, and indicate that $1 received one year from now requires that we borrow $0.91 today. In other words, $1 received one year from now is equal to $0.91 received today. Similarly, $1 received five years from now is equal to only $0.62 received today, and $1 received ten years from now is worth only $0.39 when received today. Once again, we now have a table of values that are *discounted* to adjust for the time value of money.

With the time value of money estimated, we can now review the cash flows generated from our three proposals relative to today's dollar value, and then determine which proposal generates the highest ROI and is therefore the best investment.

Table 11–4: *Discount Schedule for a $1 Loan at 10% Annual Interest*

Loan Due Date	Amount Due	Discount Factor	Inverse Discount Factor
Year 1	$1.00	1.10	0.91
Year 2	$1.00	1.21	0.83
Year 3	$1.00	1.33	0.75
Year 4	$1.00	1.46	0.68
Year 5	$1.00	1.61	0.62
Year 6	$1.00	1.77	0.56
Year 7	$1.00	1.95	0.51
Year 8	$1.00	2.14	0.47
Year 9	$1.00	2.36	0.42
Year 10	$1.00	2.59	0.39

Using the Net Present Value Method

The Net Present Value Method simply uses the discounting technique illustrated in the previous section to adjust each year's cash flows to account for the time value of money. Once all the cash flows of a given proposal are properly discounted, the discounted cash flows can then be totaled to produce a single Net Present Value (NPV) for the given proposal. Then, by comparing the NPVs of all the proposals, we can conclude which proposal is the best investment.

Note: Proposals with an NPV greater than zero are good candidates for funding because the expected returns exceed the cost of borrowed funds. Proposals that have an NPV less than or equal to zero are generally not funded because the cost of borrowed funds exceeds the expected returns. The proposal with the highest NPV provides the highest rate of return and represents the best investment.

When we apply the inverse discount factors from Table 11–4 to the original cash flows from Table 11–1 and total the results, a useful perspective emerges from the results, as shown in Tables 11–5, 11–6, and 11–7.

Table 11–5: *Net Present Value Calculations for Proposal A*

Year	Cash Flow	Inverse Discount Factor	Discounted Cash Flow
0	-$1,000	1.00	-1,000
1	500	0.91	455
2	400	0.83	331
3	300	0.75	225
4	100	0.68	68
5	0	0.62	0
Total	$300		$79

Table 11–6: *Net Present Value Calculations for Proposal B*

Year	Cash Flow	Inverse Discount Factor	Discounted Cash Flow
0	-$1,000	1.00	-1,000
1	100	0.91	91
2	200	0.83	165
3	300	0.75	225
4	400	0.68	273
5	500	0.62	310
Total	$500		$65

The three tables show that Proposal A has an NPV of $79, while Proposal B has a slightly lower NPV at $65, and Proposal C has the lowest NPV at $56. From these results we can make the following conclusions:

1. All three proposals are a good investment because they produce a positive (i.e., greater than zero) NPV. This means that each proposal generates income, even if you borrow the $1,000 at a cost of 10% interest annually.

2. Proposal A is slightly more financially attractive than the other two proposals. Therefore, if borrowing capacity is limited, Proposal A should be funded first, then Proposal B if funds are still available, and finally Proposal C.

Table 11–7: *Net Present Value Calculations for Proposal C*

Year	Cash Flow	Inverse Discount Factor	Discounted Cash Flow
0	-$1,000	1.00	–1,000
1	0	0.91	0
2	0	0.83	0
3	0	0.75	0
4	0	0.68	0
5	1,700	0.62	1,056
Total	$700		$56

Notice how much more complete our NPV financial perspective of the three proposals is compared to when we used the simpler Payback Method.

Using th Internal Rate of Return Method

While the NPV shows the actual amount of money a given proposal will generate, some firms prefer to see the results in terms of annual interest rates. To accommodate this view, financial analysts developed the Internal Rate of Return (IRR) Method, which is an alternative valuation method that produces results in terms of return percentages rather than in NPV dollars.

One simple technique used for determining IRR is to use a trial-and-error method to estimate a given proposal's annual percentage rate, where the IRR is defined as the interest rate that results in an NPV equal to zero. To accomplish this, you can use a spreadsheet application such as Microsoft Excel. By creating a spreadsheet that includes the proper formulas to calculate the discount factors, as well as the discounted cash flows generated by each proposal, and then summing the cash flow results to calculate the NPV, you can apply different interest rates until one results in an NPV of zero.

When you use the trial-and-error method to estimate a given proposal's annual percentage rate, follow these guidelines:

- If the NPV is *greater* than zero, then the given proposal's IRR is *higher* than the current interest rate used to calculate the NPV. In this case you must recalculate the NPV using a *higher* interest rate.

- If the NPV is *less* than zero, then the given proposal's IRR is *lower* than the current interest rate used to calculate the NPV. In this case you must recalculate the NPV using a *lower* interest rate.

- If the NPV is equal to zero, then you have determined the given proposal's IRR, which is quite simply the interest rate needed to produce an NPV equal to zero.

By developing the spreadsheet to calculate each proposal's NPV, you can conveniently use the trial-and-error method to estimate the IRR for the three proposals. Figure 11–1 shows the NPV results using an interest rate of 10%. Notice that each proposal's NPV matches those in Tables 11–5, 11–6, and 11–7.

To determine the annual interest rate for Proposal A, use the trial-and-error method and enter different interest rates until Proposal A's NPV equals zero. When you do so, it is determined that an interest rate of 14.5% produces an NPV of zero, as shown in Figure 11–2.

To determine the annual interest rate for Proposal B, because a 14.5% interest rate produces an NPV *lower* than zero (see Figure 11–2), we must enter a lower interest rate. Using the trial-and-error method to enter lower interest rates, it is determined that an interest rate of 12.0% produces an NPV of zero, as shown in Figure 11–3.

To determine the annual interest rate for Proposal C, because a 12.0% interest rate produces an NPV *lower* than zero (see Figure 11–3), we must enter a lower interest

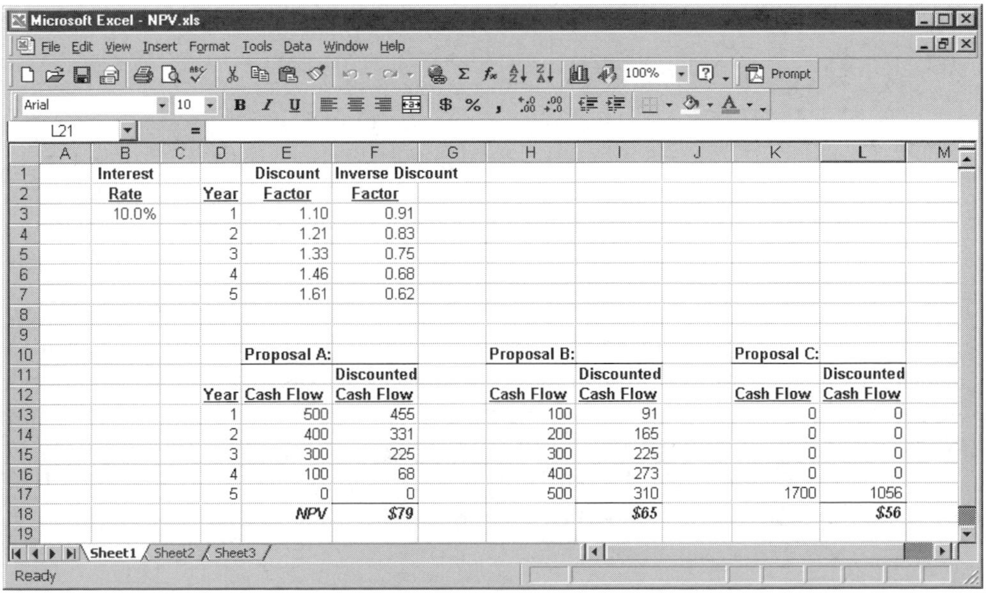

Figure 11–1 *The NPV values for the three proposals are calculated using an interest rate of 10%.*

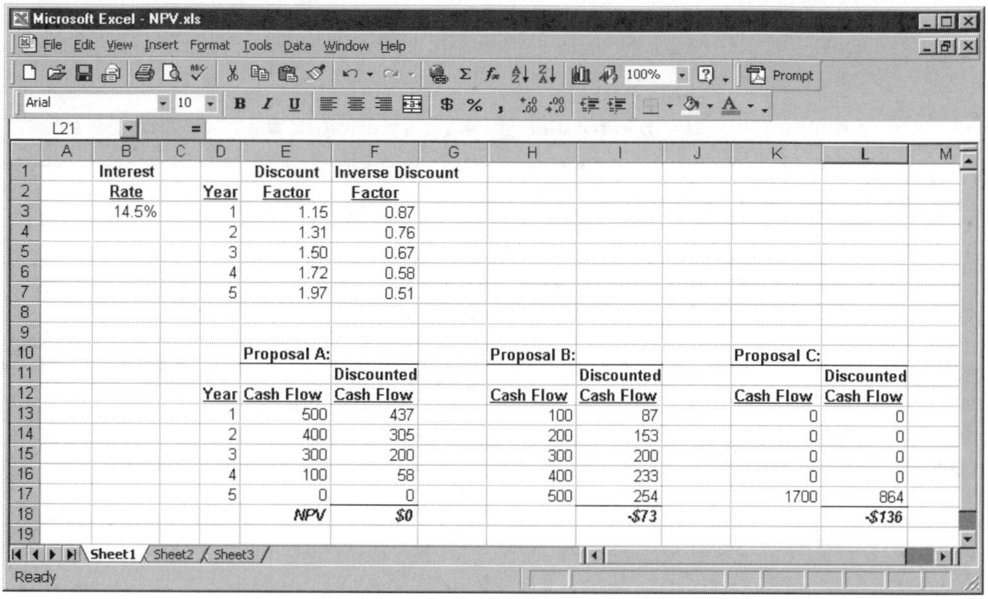

Figure 11–2 *An interest rate of 14.5% produces an NPV value of zero for Proposal A.*

The spreadsheet shown contains:

	Interest Rate		Year	Discount Factor	Inverse Discount Factor
	14.5%		1	1.15	0.87
			2	1.31	0.76
			3	1.50	0.67
			4	1.72	0.58
			5	1.97	0.51

	Proposal A:			Proposal B:			Proposal C:	
Year	Cash Flow	Discounted Cash Flow		Cash Flow	Discounted Cash Flow		Cash Flow	Discounted Cash Flow
1	500	437		100	87		0	0
2	400	305		200	153		0	0
3	300	200		300	200		0	0
4	100	58		400	233		0	0
5	0	0		500	254		1700	864
	NPV	$0			-$73			-$136

rate. Using the trial-and-error method to enter lower interest rates, it is determined that an interest rate of 11.2% produces an NPV of zero, as shown in Figure 11–4.

From the trial-and-error analysis, it is shown that a discount factor (or interest rate) of 14.5% is required to generate an NPV equal to zero for Proposal A, while a lower rate of 12.0% is needed for Proposal B, and an even lower rate of 11.2% is needed for Proposal C. In financial terms, Proposals A, B, and C are said to have IRRs of 14.5%, 12.0%, and 11.2%, respectively. Once again, this suggests that Proposal A is slightly more attractive financially than the other two Proposals, because Proposal A has a higher interest rate—the same conclusion reached by using the NPV method discussed previously.

While the IRR Method indicates which proposal generates the highest interest rate, relying on the IRR alone can lead to poor financial decisions. For example, consider a choice between investing $1 to receive $5 in one year versus investing $1 million to receive $2 million in one year. While the first choice has a much higher IRR, the second choice generates a much higher NPV. Recognizing this issue, most sophisticated firms calculate both NPV *and* IRR, but rely more on the NPV Method to determine which proposal is truly the best choice.

Note: The NPV.xls Excel spreadsheet is included on the accompanying CD.

Note: The examples and calculations used in this introduction to capital budgeting analysis are simplified to illustrate the basic principles involved. In most business situations, interest rates are compounded continuously and payments are made on a monthly basis. Also, the calculation of the "true" marginal cost of capital is significantly more complex than simply determining the prevailing lending rate at the local bank. However, these additional complexities neither change nor conflict with the fundamental lessons covered in this section.

	A	B	C	D	E	F	G	H	I	J	K	L	M
1		Interest			Discount	Inverse Discount							
2		Rate		Year	Factor	Factor							
3		12.0%		1	1.12	0.89							
4				2	1.25	0.80							
5				3	1.40	0.71							
6				4	1.57	0.64							
7				5	1.76	0.57							
8													
9													
10					Proposal A:			Proposal B:			Proposal C:		
11						Discounted			Discounted			Discounted	
12				Year	Cash Flow	Cash Flow		Cash Flow	Cash Flow		Cash Flow	Cash Flow	
13				1	500	446		100	89		0	0	
14				2	400	319		200	159		0	0	
15				3	300	214		300	214		0	0	
16				4	100	64		400	254		0	0	
17				5	0	0		500	284		1700	965	
18					NPV	$42			$0			-$35	
19													

Figure 11–3 *An interest rate of 12.0% produces an NPV value of zero for Proposal B.*

	A	B	C	D	E	F	G	H	I	J	K	L	M
1		Interest			Discount	Inverse Discount							
2		Rate		Year	Factor	Factor							
3		11.2%		1	1.11	0.90							
4				2	1.24	0.81							
5				3	1.38	0.73							
6				4	1.53	0.65							
7				5	1.70	0.59							
8													
9													
10					Proposal A:			Proposal B:			Proposal C:		
11						Discounted			Discounted			Discounted	
12				Year	Cash Flow	Cash Flow		Cash Flow	Cash Flow		Cash Flow	Cash Flow	
13				1	500	450		100	90		0	0	
14				2	400	323		200	162		0	0	
15				3	300	218		300	218		0	0	
16				4	100	65		400	262		0	0	
17				5	0	0		500	294		1700	1000	
18					NPV	$57			$26			$0	
19													

Figure 11–4 *An interest rate of 11.2% produces an NPV value of zero for Proposal C.*

SUMMARY

In this chapter you learned how to communicate better with upper management by speaking their language, the language of finances. By calculating the Return on Investment (ROI) of CAD-related expenditures such as software and training, you effectively communicate the company's needs to those responsible for generating a profit. Additionally, by using the methods described in the Introduction to Capital Budgeting Analysis section, you can better grasp which of your proposals—your requests for funds for software and training—have a good chance for approval. Through discussion of the tools and techniques covered in this chapter, you have gained insight into the challenges facing upper management and are better prepared to help them understand your needs.

Developing CAD Standards

CAD standards function as both the table manners and the rules of the road for creating CAD drawings. If you are a CAD manager, then you will probably spend some of your time playing mom and traffic cop as you develop, maintain, and enforce CAD standards.

CAD standards are like table manners in that they make it easier for groups of people to approach a task in a similar way, without forcing everyone to make decisions about minutia or letting unusual behavior disrupt the proceedings. When a group of people knows which forks and wine glasses to use for which purposes, they can concentrate on the more important task of enjoying the meal and each other's conversation. When AutoCAD users know which layers and file names to use for which purposes, they can focus on the more important task of creating accurate, good-looking drawings.

CAD standards are like rules of the road in that they prevent disastrous collisions when lots of people need to share the same drawings. When drivers use turn signals and yield at the right time, they are less likely to run into each other and more likely to get to their destinations quickly. When CAD users on a project conform with a consistent set of CAD standards, they are less likely to run into problems sharing each other's drawings and more likely to finish those drawings on time.

Developing, maintaining, and enforcing CAD standards is not an easy job. It is always a struggle between the ideal goal of doing a thorough, careful job and both the practicalities of squeezing CAD standards work into busy production schedules and the demands to maximize billable hours. This chapter defines the CAD standards development process and describes practical techniques and tools that make the job easier.

CAD STANDARDS: WHAT AND WHY?

CAD standards are the offspring of the graphical standards from manual drafting days: how tall text should be, how to depict the parts of the real-world objects that you are representing, what dimensions and drafting symbols should look like, and so forth. If you work in any of the AEC (architecture, engineering, and construction) professions, then you will probably find in the office somewhere an old copy of *Architectural Graphic Standards* that addresses these kinds of questions. In any case, the emphasis in manual drafting standards was always on achieving graphical consistency in all the drawings.

CAD standards aim to achieve the same thing—graphical consistency on plotted drawings—but in addition they address a range of new issues that were brought on by computers and CAD software. Chief among these issues are the naming and use of layers and files. Other issues include drawing setup and organization (using paper space and/or xrefs), text and dimension styles, and plotting standards (especially color-to-lineweight mapping).

FIGHTING INCONSISTENCY

Perhaps the most important reason to create a CAD manual is that the process forces you and your company to grapple with the CAD standards monster. Inconsistency in a company's CAD drawings often is not the fault of users not following the rules, but is due to there not being any rules (or of there being obsolete or impractical rules).

The inadequacy of many CAD manuals is a result of their being little more than dusted-off versions of old manual drafting standards. These old standards were concerned with the look of the drawing on vellum or mylar. The means of achieving that look was not addressed in most manual drafting standards because the techniques and tools of the trade of manual drafting were relatively consistent, few in number, and part of the training of most drafters. CAD changed all that; now there are a wide range of CAD programs and companion applications, an even wider range of techniques for using them, and a much broader spread of skill levels among CAD users. The look of the plotted drawing remains important of course, but how each user organizes the drawings and objects in them is equally significant. CAD standards thus need to address the process as well as the result.

In developing or revising a CAD manual, you create the opportunity to question broadly how CAD is used in your company. How are things done now? What things take a lot of time, and what things seem relatively efficient? What good ideas from various users can be spread around to everyone else? What things need to be standardized, and what things can be left to individual choice?

Raising all of these questions is not going to make your life any easier in the short run. It takes a lot of work for one person to come up with a good set of CAD standards. The complications and concomitant effort increase with the number of people affected. Easygoing AutoCAD users can become impassioned almost to the point of belligerence if their favorite way of naming layers or using paper space is not adopted as the company standard. Nonetheless, it is important to include those who are affected by the standards in their creation. Some of those people may have good ideas. More importantly, the odds of getting everyone to adhere to standards are better when people feel like they have been a part of the process.

CAD STANDARDS AS A MEANS TO AN END

CAD standards are not ends in themselves. Their purpose is to improve the efficiency of the CAD process and the quality of the end result. Specifically, CAD standards can help with the following:

- **Production efficiency**: It is usually a lot faster to create drawings if everyone is working with a prearranged, well-ordered set of layers, symbols, and other items—especially if you have customized the CAD software to support and encourage the use of those standardized items. (On the other hand, overly complicated or badly implemented CAD standards can slow down production, as users struggle to create the right items and use them correctly.)

- **Plotting consistency**: Yes, most of us still plot a lot, and when we do, we want consistent output. For example, on architectural plans, grid lines should be thin, text should be medium, and columns should be thick. All the sheets for a project should look consistent and read well. One of the keys to this consistency is a reasonably standardized set of layers and layer properties, along with using the system whereby individual objects inherit their properties from the layers that they are on (in AutoCAD terminology, color, linetype, lineweight, and plot style "by layer" rather than "by object"). Another key to plotting consistency is a standardized set of plotting support files and procedures.

- **Electronic consistency**: The consistency of drawing data inside the CAD file is increasingly as important as the consistency of plotted output. In fact, electronic consistency can be more important, especially if you use custom applications that manipulate or extract data from your CAD drawings.

- **Ease of reuse by other people or companies (electronic drawing exchange)**: In many CAD-using industries, different companies that collaborate on projects frequently make use of each other's drawings, often as "backgrounds." For example, a plumbing drawing might use the architectural or structural drawings as backgrounds. In many cases, one profession wants to show only some of the objects that appear on the other professions' full drawings. When layer use is inconsistent, this kind of "partial reuse" becomes extremely difficult.

CAD standards will not magically make these things appear, but they do provide a framework within which talented and intelligent CAD users can bring about the desired benefits.

ANATOMY OF A CAD STANDARD

So what exactly should be in a CAD standard? For many companies, it used to be as simple as a list of layer names and a color-to-lineweight plotting chart, sometimes supplemented by the company's old manual drafting graphical standards. Most companies now realize that CAD standards need to cover a much wider range of CAD practices and organizational techniques. The contents and organization of each company's CAD standard will vary depending on discipline, work specialty, sophistication, and other factors, but the following topics indicate the contents and scope of a well-rounded CAD standard:

- **Drawing setup and drawing set organization**: Use of xrefs; use of paper space; use of blocks and attributes; title blocks; model space and paper space setup procedures

- **File and folder management**: Folder structure, naming, and contents; file naming; rules about xrefs in different folders

- **Layers**: Names of and what goes on each layer; default properties for layers (color, linetype, lineweight, etc.); rules about applying properties directly to objects (rather than leaving all properties by layer)

- **Annotation**: Text (including names of styles, uses of styles, and text heights); dimensions (including names and uses of styles); hatching (patterns, scales, and angles); symbols (section marks, callouts, tags, etc.)

- **Common drafting procedures**: Different kinds of drawings/models/sheets; mixed-scale sheets; accuracy and precision requirements; drawing revisions

- **Plotting**: Color-to-lineweight mapping; plotting procedures (use of PC3 files, CTB files, STB files, etc.)

- **Drawing exchange**: File-packaging procedure; documentation requirements; sending and receiving procedures

Your industry or office may have additional CAD standards requirements. You can research this possibility by asking others in your profession and company. What kinds of things are important to standardize? What things historically have *not* been shown consistently on the company's drawings? What are project managers or clients griping about on your company's drawings? CAD users' groups are a great source of advice about CAD standards. CAD managers often frequent these groups, and many of them have experience developing CAD standards in different industries.

A CAD STANDARDS HIERARCHY: INDUSTRY, COMPANY, PROJECT

The wag who once pointed out that "the wonderful thing about standards is that there are so many to choose from" might well have been talking about CAD standards. Ideally, each industry would have its own CAD standard (a real *standard!*), that would simplify exchanging drawings among companies in that industry and greatly reduce each company's CAD standards development work. Unfortunately, no CAD-using industry has settled on a consistent way of naming files and layers, never mind the more complex questions of how to organize drawings using tools such as xrefs and paper space.

The news is not all depressing though. CAD sophistication has increased in all industries and in most companies. The increasing frequency of electronic drawing exchange is driving many companies to pay more attention to CAD standards. Increased drawing exchange means that more people see more examples of CAD practice, and the better methods are slowly driving out the worse ones. As a result, many companies are converging on a few proven methods. It would be overstating the case to claim that there is anything close to unanimity within any industry, but at least in some industries the CAD standards story is not quite the Tower of Babel that it used to be.

(In the United States, many AEC firms have settled on company CAD standards that are based at least loosely on *CAD Layer Guidelines*, the second edition of which was published in 1997 by the American Institute of Architects (AIA). There is lots of variation in how different companies interpret, extend, and depart from these guidelines, but at least there has been moderately widespread adoption of a consistent general approach to naming layers—and, to a lesser extent, files.)

In the absence of industry standards, most companies spend some effort developing, documenting, maintaining, and enforcing internal, company-specific CAD standards. Those efforts help improve consistency within companies, even if they do not directly solve the wider industry problem.

The Perils of Project-Specific Standards

People who spend a lot of time developing CAD standards sometimes begin to suffer from the delusion that theirs is the only or best way. If these people work for companies that are in the position of directing subcontractors, they may then feel compelled to inflict their standards on everyone in the subcontractor companies.

Part of the problem is that the people in a company who write and negotiate contracts usually are not very familiar with CAD and thus are not aware of how time-consuming and expensive it is to conform with project-specific standards. The person at the contracting company figures that CAD standards must be something like sheet numbering, and of course they dictate it. The person at the contracted company figures that because CAD is magic, the drafters will just flip a switch somewhere on their computers. And besides, "the client is always right," right?

In reality, it is tremendously disruptive to switch to a completely different CAD standard that was developed in a different office with different working styles, companion applications, customization, training, and so on. It is even more disruptive to switch among several completely different CAD standards as you switch among projects! You have to reconfigure companion applications and custom utilities to support a different standard and acquaint users with the differences. CAD standards can never be completely automated, so drafters have to be trained, reminded, and browbeaten. (Of course, clients have a right to demand, "do it our way, or you don't work with us," especially on larger projects. But those clients who make such demands need to face up to the costs of doing so and augment their subcontractors' compensation accordingly.)

Drawing Exchange Guidelines

On the other hand, drawing exchange *guidelines* are important. At the very least, companies that work together on a project should agree to use naming schemes for files, layers, blocks, and text and dimension styles that avoid conflicts. It is extremely helpful to agree on a project base point so that drawings from different companies line up when they are xref-ed or inserted at (0,0).

The specifics of layers (beyond avoiding name conflicts) can be trickier. If companies want to reuse data from other companies' drawings, then each company needs to be internally *consistent* about using layers. In addition, each company may need to reconsider layer *granularity* (which things belong on the same layer and which things need to be on separate layers so that other companies can isolate them). For example, the structural drafter would prefer that bearing walls and non-bearing partition walls were on different layers in the architectural plans. Consistency and granularity of layer use are much more important than layer names, because the names are relatively easy to convert automatically using the AutoCAD 2002 Layer Translator utility. (To open the Layer Translator dialog box, shown in Figure 12-1, enter LAYTRANS at the command prompt or from the Tools menu, choose CAD Standards>Layer Translator).

Avoiding naming conflicts and maintaining reasonable layer consistency and granularity really should be worked out in each company's office CAD standards, rather than on an ad hoc, project-specific basis. If you do get these matters sorted out within your company, then project CAD planning can focus on real, project-specific issues such as base points and any unique aspects of the project that require special coordination.

If you find yourself on the receiving end of project-specific CAD standards (and you probably will), you have two choices: (1) Change how you work throughout the project, or (2) continue doing things in your customary way and convert the drawings before sending them out. Each approach has pluses and minuses. Which is the

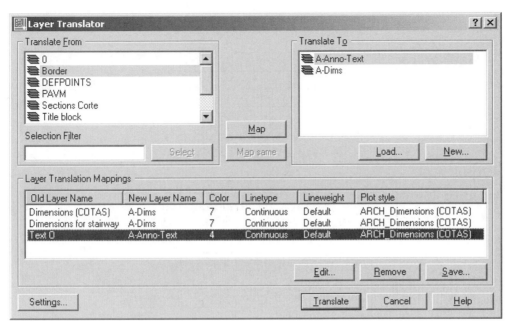

Figure 12–1 *Layer Translator dialog box.*

lesser of the two evils will depend on the specifics of the project-specific standards, how often you exchange drawings, how agile your CAD users are, and how tolerant you client is.

THE CAD STANDARDS DEVELOPMENT PROCESS

Developing company CAD standards is not most people's idea of fun. It is not fun for you because it involves lots of unpleasant tasks: reading boring standards documents written by bureaucratic committees, listening politely to (sometimes foolish) suggestions from bosses and users, and writing and editing a document that's long enough to cover most of the bases but short enough so that someone actually will read it. It is not fun for your company because it takes you away from other work, most of which is probably billable to clients.

Even worse, the job is never done. New versions of CAD programs, industry changes, revised office practices, and the inexorable march of technology ensure that your CAD standards are never up to date. Figure 12-2 maps out one version of the process, complete with endless loops.

DEFINE GOALS AND PRIORITIES

Before you dive into creating layer charts and lists of drafting rules, you should take a moment to define the goals and priorities for CAD standards in your office. What are the basic CAD approaches that your CAD standards try to reflect? (For exam-

The CAD Standards Development Process

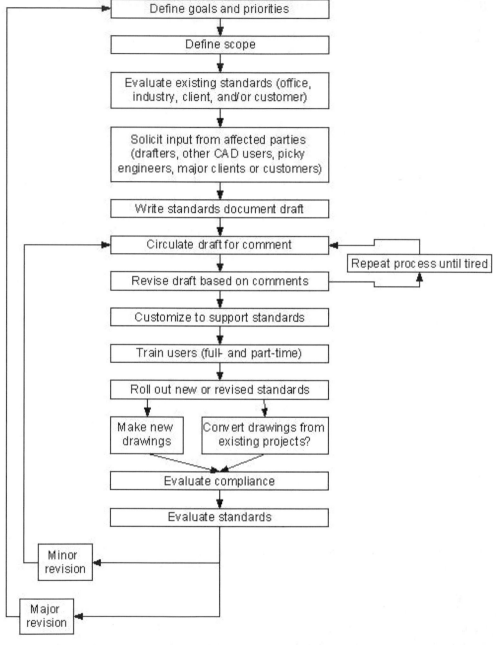

Figure 12–2 *The CAD standards development process.*

ple, "We focus on plotted drawing presentation and don't worry too much about electronic structure", or "We're really picky about using xrefs and paper space in this way.") Companies often expect CAD standards to do everything: improve production efficiency, increase precision, enhance plotted drawing quality, enable better data extraction from drawings, and make all clients and subconsultants happy. You will be more successful if you can articulate which are the most important goals and aim your standards development at them.

In addition, there are situations in which there is tension between two or more of the goals. (For example, "We know that this hand-lettered TrueType font slows down AutoCAD performance, but we really like the 'manual drafted' look that it gives to our drawings.") It is useful to have an agreed-upon criterion for handling these situations.

DEFINE SCOPE

After you have established goals and priorities, take another moment to define the scope of your current standards development effort. Because the job is never done, you should focus on the most important goals and leave lower priorities times for the next pass. Defining a reasonable scope not only helps you, it also increases the odds that users will be able to digest the new or revised standards when you roll them out. The list in the "Anatomy of a CAD Standard" section earlier in this chapter provides suggestions for topics that you might want to include in your scope.

As part of your scope definition process, you should ask for whom you are developing the CAD standards. Just the full-time CAD drafters? Part-time CAD users such as architects and engineers? Clients or consultants? The larger the circle of users or beneficiaries is, the more effort you will have to put into developing and maintaining your CAD standards.

EVALUATE EXISTING STANDARDS AND CUSTOMS

In most cases you will not be developing CAD standards in a complete vacuum. You will often start from existing company CAD standards, however incomplete or out of date they might be. Even if there are no documented company standards, there may be common practices that a majority of CAD users have been employing.

As was mentioned earlier in this chapter, there may standards documents or de facto CAD practices for your industry. Whether or not you adopt these, you should review them for ideas. If a documented standard seems mostly sensible (and/or has been embraced by a lot of companies in your industry), then you should seriously consider adopting it, at least as a starting point. Doing so usually will save you a lot of CAD standards development time.

GET INPUT FROM AFFECTED PARTIES

In the early days of CAD, CAD managers often got to create CAD standards on their own, with little meddling from others. Few principles or project managers used

CAD, so they did not care what the CAD standards said, as long as the plotted drawings looked more or less like the old pencil or ink drawings created by the office's manual drafting patriarch.

Nowadays most people do recognize the importance of CAD standards, and a lot of them want to have a say in it. We have also become more sophisticated about how we create drawings and have grown to depend more on the electronic exchange of drawing files. As a result, the process of developing CAD standards has grown much more complicated and usually involves far more people.

When you are developing CAD standards, it is generally a good idea to involve the people who will be affected by them in the development process. The affected parties might include the following:

- **Full-time CAD users**: These are likely to be the people who are most affected by CAD standards decisions, and who have the strongest opinions about them!

- **Occasional CAD users**: More people with a variety of levels of CAD experience are getting their hands on CAD files. Working on CAD drawings is no longer the sole province of a priesthood of full-time CAD drafters. In many companies, engineers and architects are opening drawings in order to view, plot, and even edit them. Many of these part-time CAD users are less attuned to the importance of CAD standards and how to abide by them when they plot or edit. Also, they are more likely to use lower-cost CAD programs such as AutoCAD LT, which provide limited support for customization that automates standards compliance. As more of these part-time CAD users touch more drawings, it is going to become essential to bring them into the CAD standards fold.

- **Company managers who do not use CAD**: Managers usually will care more about traditional "how the drawings look" issues and about the effect of CAD standards decisions on productivity.

- **Other offices of a large company**: If you are developing CAD standards for one office in a larger company, then you will need to consider how your standards mesh with (or do not mesh with) standards in the other offices.

- **Clients**: Your clients may care about your CAD standards, especially if you exchange drawing files with them frequently. If you have major clients that you work with often, you will probably want to consider how to keep them happy.

- **Subconsultants**: Similarly, subconsultants that you work with may have an interest in your CAD standards for drawing exchange reasons. In this case, you are in the position of power, but you still might want to consider a subconsultant's needs, because they can influence the efficiency and smoothness of work on mutual projects.

Of course, you have to balance involvement by a wide range of affected parties against creating an efficient CAD standards development process. If too many people are actively involved, the development effort will get bogged down, and you will never finish. The key is to define hierarchies of affected parties and give each group the opportunity to provide an appropriate amount of input. For example, you might want to involve full-time CAD drafters—who are the most directly affected by CAD standards—in meetings where options are discussed and decisions are made. On the other hand, you might want to limit part-time CAD users to commenting on draft versions that already contain most of the decisions.

MAKE DECISIONS

In many cases, decision making is the easiest and quickest part of the process. In a few instances, though, it is likely to be difficult to reach consensus on an important question. For this reason, it is useful to agree ahead of time on a procedure for getting beyond impasses. You might have a voting procedure (such as "majority rules") or designate a person whom everyone respects as the final decision maker for contentious questions.

DOCUMENT DECISIONS

The development of the actual standards involves creating some sort of document, usually in a word processor format, Windows Help file, or HTML (i.e., Web pages). Assuming that everyone in your office is networked and Web-browsered, HTML is an excellent way to go. Most users will refer to Web-based CAD standards documentation more frequently than printed documentation, and obviously it is much easier to update HTML documentation and ensure that everyone is looking at the current version.

Usually you will create a draft document, circulate it for comments, and then revise it based on the comments. This cycle can go on forever, so at some point you have to call it "good" and move on to the implementation phase. It helps to have a clear chain of command that ends with someone who has the responsibility and the clout to say, "Okay, we're done with this part."

The scope and format of your CAD standards documentation should reflect the realities of how it will be used:

1. It will guide your customization of AutoCAD and your companion applications and then reflect the fully customized system.
2. It will be an orientation document for new hires, temporary employees, or subcontractors whom you expect to follow your CAD standards.
3. It will be a reference document for CAD users when they have standards questions.

These uses suggest that a good CAD manual should be relatively short and easy to browse. Anything longer than a few dozen pages will not get read, especially by a temporary employee or subcontractor. A competent AutoCAD user should be able to skim the document quickly and come away with a pretty good understanding of how your company uses CAD. Users should be able to find things quickly (via a table of contents and/or index) and the layout should support at-a-glance comprehension (tables, bullet lists, figures, screen captures, etc.).

The usual CAD manual problems are matters of too little or too much. Companies that have not spent much time on CAD standards development often have no more than a few pages of layer names and maybe a plotting chart showing color-to-line-weight correspondence. That is a good start, but it neglects many of the important standards issues mentioned earlier. The opposite sin is an enormous three-ring binder that becomes the CAD manager's vanity publishing effort. If you want to write a book, fine, but a CAD manual is not the place for that much text. More is not always better, and a CAD manual that runs into hundreds of pages is doomed to irrelevance.

Some companies also include CAD system information—such as descriptions of hardware and software, list of keyboard shortcuts, and custom program documentation—in their CAD standards. While this information certainly is useful, it tends to bloat the CAD standards document. If you want to document these kinds of things, consider putting them in a separate CAD systems manual.

You will want to organize your CAD standards document into chapters or other logical subdivisions, perhaps along the lines of the list in the "Anatomy of a CAD Standard" section earlier in this chapter. In addition, consider providing these sections in order to make the document easier to use:

- Table of contents
- Revision history
- Introduction (including purpose and scope)
- Overview
- Index

A CAD standards overview is particularly useful as an "executive summary" for managers, temporary hires, and people in other companies with whom you exchange drawings. The overview should cover guiding principles (standards priorities, conformity with any industry standards, overall file- and layer-naming scheme etc.); major working methodologies (drawing organization, xrefs, paper space); and major "thou shalt/thou shalt not" rules (e.g., no overriding of "by layer" object properties).

CUSTOMIZE TO SUPPORT STANDARDS

A surprising number of CAD managers continue to believe, despite the verities of human nature and all evidence to the contrary, that every CAD user will read the company CAD manual cover to cover (carefully, several times), and then refer to it continually. In fact, you are lucky if most of them thumb through all of your hard work once.

If you rely on a CAD standards document alone—whether it is printed or Web-based—standards compliance inevitably will be erratic. Users will continue to do things the way they are used to, either deliberately ("my way is easier") or through ignorance ("I didn't remember that part of the manual"). In order to thwart these two problems, you must use AutoCAD customization to automate much of the work of standards compliance. At a minimum, your customized system should make it no more onerous to create drawings according to company standards than to create them without any standards. Better yet, the CAD system should make it *more* efficient to draw according to company standards than not to, by automating layer changes and other repetitive tasks.

Thus the more realistic and efficient means of enforcement is to build as many of your standards as possible into your CAD system. Most companion applications include mechanisms for configuring the layers, text styles, and other items that the application uses. These settings should ensure that objects created with companion application commands are drawn according to your company standards—without extra steps or guessing by users. Application settings will not help when users employ ordinary AutoCAD commands (e.g., LINE or MTEXT) to draw objects. For those cases, you might want to add menu macros, scripts, or AutoLISP or VBA programs to streamline layer creation and other tasks.

You will never completely automate standards enforcement or eliminate the need for trained users who understand what they are doing. But a combination of companion application configuration and additional AutoCAD customization can cover a large majority of your CAD users' needs.

The customization tools in AutoCAD that can be helpful for automating standards compliance include the following (see Parts 3 and 4 of this book for more information):

- Template drawings
- Blocks
- Scripts
- Custom pull-down menus and image tile menus
- Custom toolbars
- Plotting support files: PC3, PMP, CTB, STB
- AutoLISP
- VBA (Visual Basic for Applications)

You can use these tools to create things like drawing setup utilities, standard text and dimension styles, macros that automatically set layers, symbol libraries, and stream-lined plotting procedures. If people in your company use both AutoCAD and AutoCAD LT, bear in mind that any tools you develop with AutoLISP or VBA will not be accessible to the LT users.

Here are some specific suggestions for how to customize to support CAD standards:

- Create template drawings containing the system variable settings, layers, text styles, and dimension styles that most of your drawings require.

- Create drawings that contain groups of standard layers, text styles, dimension styles, and symbols (i.e., blocks). Users can insert these drawings as blocks in order to "populate" their current drawing with the named objects. Alternatively, they can use AutoCAD DesignCenter to "mine" the drawings for the named objects that they want.

- Create scripts that make groups of standard layers, along with their standard properties—color, linetype, and so on. (Note that the script method of creating layers, unlike the previous two methods, will correct colors, linetypes, and other properties that users might have changed.)

- Create custom pull-down menus and image tile menus that insert standard symbols.

- Create custom pull-down menus and toolbars that load and run AutoLISP, VBA, and ARX programs (some of which might draw objects that conform to your CAD standards).

- Create plotting support files that contain the proper settings (e.g., color-to-lineweight mapping) for plotting according to your standards. Attach the appropriate PC3 and CTB (or STB) files to each folder in your template drawings so that users start out with sensible, standard plot settings.

- Create AutoLISP or VBA programs that perform drawing setup tasks, draw objects that conform to your CAD standards, or that help users in other ways to observe standards.

In many cases you will have the choice of customizing a particular CAD standards item by creating it in a file or writing a program that creates it on the fly. For example, you could customize dimension styles, text styles, and drawing scale settings by putting them in a series of drawing template files (one for each drawing scale that you use). Alternatively, you could write a single AutoLISP or VBA drawing setup program that creates the styles and settings when the user sets up a new drawing. The programmatic way is almost always more elegant. It requires fewer files; thus there are fewer places to make changes when you revise your CAD standards. Also, programs usually are much more flexible than static files. On the other hand, writing a program requires more sophistication and may take more time than creating a few static files.

TRAIN AFFECTED PARTIES

Although customization to support standards is vital, do not get caught in the trap of thinking that customization by itself will magically force all users to follow standards all the time. There is no such thing as an idiot-proof CAD system or idiot-proof standards customization. Users must be trained, and that means *all* users, including the ones who occasionally open drawings to plot, view, or do light editing.

It is best to schedule an orientation session at the rollout of new or revised CAD standards. Walk everyone through the major points and most significant changes. Point out custom tools that will help with standards compliance (and also improve efficiency). Most people are not going to absorb all the standards in a single session. One good approach is to weave follow-up standards training into a regular training regimen.

Do not forget about temporary or new hires who show up after the initial implementation of new CAD standards. You should have an orientation process for these people so that they can maintain at least reasonable consistency with what other people are doing. If you use temps, it pays to give some thought to how to orient them quickly, because there is rarely time for a full orientation. As mentioned earlier in this chapter, a short CAD standards overview in your standards document will help get temporary people quickly up to speed with the most important things.

ROLL OUT NEW STANDARDS

With the CAD standards documentation, customization, and some training in place, you can "roll out" the new standards officially. It is a good idea to let everyone in the company know—even those who do not use CAD—so that there is not confusion among managers or others who work with CAD users.

One sticky problem when you revise your CAD standards is what to do about current projects and existing drawings that were created before the new standards went into effect. Your choices are to continue doing things the old way on old projects or convert the drawings so that they conform to the new standards. Your decision usually will depend on how far along the project is and how much work would be involved in conversion.

A similar challenge arises with block libraries, typical details, and the like. You will almost certainly want to convert these, but you may want to do it on a copy of the files, so that you have the older versions around for use with current, unconverted projects.

"Manual" conversion of drawings from an old CAD standard (or no CAD standard) to a new CAD standard is tedious at best. Use the AutoCAD 2002 Layer Translator, described earlier in this chapter, and the Standards Manager, described later in this chapter to help automate the conversion process.

EVALUATE STANDARDS AND COMPLIANCE

At some point after the new CAD standards have been in use for awhile, you should take a look at how well they are working:

- When people do follow the standards, do they give the desired results? (In other words, are the standards achieving the goals and priorities that you established in the beginning of the standards development process?)

- How well in fact are people following the standards? (See the next section for more information about standards enforcement.)

You will use the answers to these questions during your next revision to refine the contents, approach, and compliance mechanisms of your CAD standards.

STANDARDS ENFORCEMENT

With your new CAD standards in place, the real fun begins—trying to get everyone in the company to comply with them. The training and customization efforts described earlier in this chapter will help, but if compliance is important to you, then you will need to do some policing too.

One of the great conundrums of CAD standards is that there is no straightforward, economical way to verify compliance of every drawing with every single rule. The AutoCAD 2002 CAD Standards tools (described later in this section) do a lot to enhance your compliance-checking abilities, but they do not check everything. Thus any CAD standards enforcement scheme will, out of necessity, be only partially successful. A good enforcement scheme usually combines the carrot (customization, rational explanations, appeals to enlightened self-interest, praise for good compliance) and the stick (complaints, threats, punishment).

In many companies, checking for compliance is something that the CAD manager or users do in the course of working with drawings from other people. Someone opens a drawing and notices that the layer names are a mess or that the drawing was set up incorrectly. In an office with lots of sharing of drawings and a sufficient number of people who care about standards, this kind of informal checking may be enough to keep compliance at a reasonable level.

Some companies implement a periodic drawing review process. For example, the CAD manager or a project manager might request a drawing review, which consists of someone checking the drawings using a checklist that is part of the company's CAD standards document.

Your evaluation of compliance also acts as an indirect check on the reasonableness of your CAD standards and the thoroughness of their implementation. If the same "mistakes" show up repeatedly in different drawings created by different people, that is a good sign that (1) the standard is not sensible or comprehensible, (2) users need more training, and/or (3) you need to do more customization to support the standard better.

Autodesk has been focusing some of its development efforts on CAD and layer standards recently. Two of the extensions for AutoCAD 2000, Layer Translator and Standards Manager, are now part of AutoCAD 2002. Most of the CAD Standards tools are available via the Tools>CAD Standards submenu, or from the CAD Standards toolbar, which is shown in Figure 12-3

Figure 12–3 *CAD Standards toolbar.*

The Layer Translator (see Figure 12-1 earlier in this chapter) not only translates layer names from one layer system to another, it can also perform other standards-enforcement tasks automatically, such as forcing object properties to be "by layer" and purging unreferenced layers.

The Standards Manager checks layers and several other CAD standards items against a set of company standards. You can check the current drawing and, in some cases, automatically fix discrepancies. To use the Standards Manager, you first define your standards in the Configure Standards dialog box (see Figure 12-4). You then check individual drawings—and optionally fix some of the discrepancies automatically—with the Check Standards dialog box (see Figure 12-5).

Alternatively, you can batch-check a group of drawings (see Figure 12-6), but only in report mode—there is no automatic fixing in batch mode. The Batch Standards Checker is a separate program – in order to launch it, choose the Windows Start menu, then Programs, AutoCAD 2002, Batch <$startrange>CAD standards: enforcingStandards Checker.

CAD STANDARDS RESOURCES

As was mentioned earlier in this chapter, the AIA publication *CAD Layer Guidelines* has become popular as a basis for layer and file naming in many AEC companies. The second edition was published in 1997, and it received an additional boost when it was adopted by the National Institute of Building Sciences (NIBS), with some amendments, as part of its National CAD Standard (NCS). Two *CADALYST* magazine "CAD Manager" columns describe these two documents:

- "CAD Standards: The nitty-gritty" (May 1998):
  http://www.cadalyst.com/solutions/mc/598mc/598mc.htm

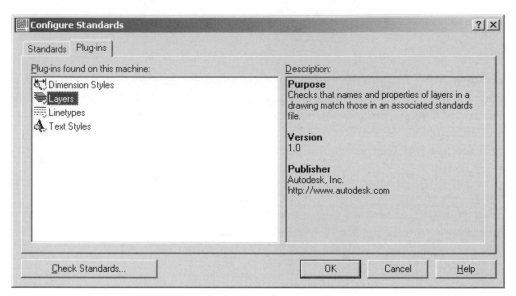

Figure 12–4 *Configure Standards dialog box.*

- "The NIBS National CAD Standards, v1" (April 2000):
  http://www.cadalyst.com/solutions/mc/0400cm/0400cm.htm

Unfortunately, neither *CAD Layer Guidelines* nor the NIBS National CAD Standard are available online—you have to buy the hardcopy versions. You can purchase *CAD Layer Guidelines* from the American Institute of Architects for about $30 (1-800-365-2724, order number R809-97). Ordering information for the National CAD Standard is available at http://www.nationalcadstandard.org.

If you would like to explore the question of layer standards further, read the *CADA-LYST* magazine article "AEC Layer Debate—AIA Sets the Standard" (at www.cadalyst.com/features/0601layer/0601layer.htm). Although the article primarily addresses people in the AEC professions, it also contains general considerations for deciding on and documenting layer standards.

The AIA's *CAD Layer Guidelines* are not the only attempt at a documented, industry-wide layer standard. Here are some others:

- The International Organization for Standardization (ISO) publishes a trio of documents covering CAD layers: ISO 13567-1, ISO 13567-2, and ISO 13567-3. (http://www.iso.ch)

- The British Standards Institution's BS 1192-5 is a popular AEC standard among CAD users in the United Kingdom. (http://www.bsi-global.com)

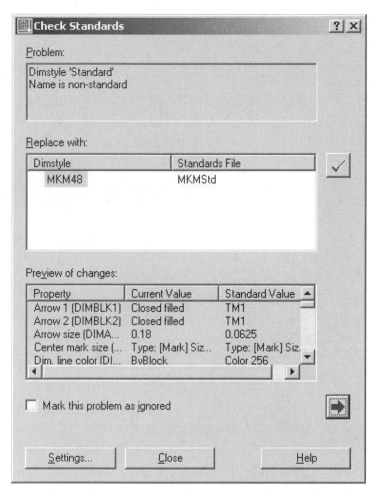

Figure 12–5 *Check Standards dialog box.*

- Softco sells the S-MAN Standards Management system—a "fill in the blanks" system for AEC drawings with translation software for converting between different sets of standards. (http://www.softcosys.com)

- CADCOM sells the ASCAD system for civil engineering drawings. (http://www.cadcom.ca)

- The Grafix Shop sells Civil Drafter, another CAD standards system for civil engineering drawings. (http://www.thegrafixshop.com)

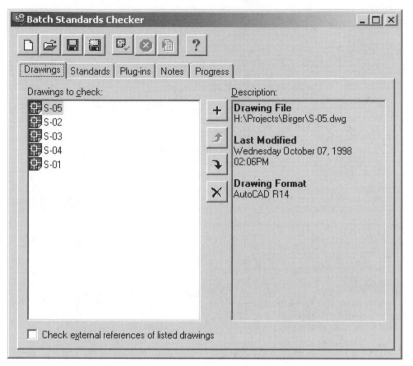

Figure 12–6 *Batch Standards Checker dialog box.*

SUMMARY

In his chapter you learned what goes into a well-rounded set of CAD standards and how to go about the unending job of developing, maintaining, and enforcing them. Accept the fact that CAD standards always will be a work in progress, and do not let "best" be the enemy of "good" (or "good" be the enemy of "not too bad," for that matter). Also, keep in mind that CAD standards should play a supporting role in the drawing creation and exchange process. CAD standards are not ends in themselves; they are the means to high-quality, efficiently produced, consistent-looking drawings.

CHAPTER 13

Managing CAD Workflow

One of the biggest challenges you will face in completing a CAD-based project is managing CAD workflow. Controlling file access, tracking revisions to drawings, and sharing drawings among CAD technicians, designers, and managers are essential tasks that if not performed well, will lead to misplaced files, duplicated work efforts, and lost productivity. By managing CAD workflow, you control your CAD-based project from its conceptual beginning to its final deliverables, and you avoid missing deadlines and blowing budgets.

In this chapter you will learn about various concepts, techniques, and tools for managing CAD workflow, including the following:

- File naming and directory standards
- Controlling file permissions
- Revision and approval control
- Viewing, printing, and drawing markup
- Wide-area access via the Internet

FILE NAMING AND DIRECTORY STANDARDS

The foundation of managing CAD workflow involves establishing file and directory (folder) standards. When you develop a systematized approach to naming your files and folders, you establish a method for adding new files and folders to your project, allowing your project to grow in an orderly, predictable fashion, and thus making your project scalable.

There are two key concepts that will help you manage this portion of CAD workflow:

- Developing an efficient drawing numbering system
- Developing a file naming/folder naming system that corresponds to the drawing numbering system

DEVELOPING AN EFFICIENT DRAWING NUMBERING SYSTEM

Drawings are the lifeblood of a CAD-based. Successful CAD workflow management requires an efficient drawing numbering system—a numbering system that is intuitive (easy to understand) and scalable (that lets you continuously add new drawings to your project.) By developing an efficient drawing numbering system, you ensure that a drawing's contents are easily identified without having to open the file, and that drawing files are easy to manage and track.

To manage CAD workflow you need a way to name or number your drawings that is consistent and that allows flexibility in organizing drawings by type (such as street improvement plans versus existing utility plans) and adding new drawings to your project. An example of an efficient, flexible numbering system is a library catalog system, more commonly known as the Dewey Decimal System.

The Dewey Decimal System organizes books into categories and subcategories. This approach makes it easy to identify a book by its category and subcategory, and it lets you add new books to the existing category sequence. If you organize your drawings into categories and then sequence the drawings within a category, this type of system might work for you. For example, in the AEC industry, drawings have categories such as Cover Sheets and General Information Sheets, Floor Plans, Elevations, and Sections. Each category might contain a sequence of drawings, such as the First Floor Plan and the Second Floor Plan. Following this format, you could define a drawing numbering system that meets the goals of predictability, flexibility, and scalability. The following list exemplifies such a numbering system:

- 0.00, 0.01, 0.02, ..., 0.*nn*—Cover Sheets and General Information
- 1.00, 1.01, 1.02, ..., 1.*nn*—Floor Plans
- 2.00, 2.01, 2.02, ..., 2.*nn*—Elevations
- 3.00, 3.01, 3.02, ..., 3.*nn*—Sections
- *n*.00, *n*.01, *n*.02, ..., *n*.*nn*—Additional categories

Using this approach to organize drawings, a new floor plan drawing may be added to the end of the Floor Plans category at any time without adversely affecting the drawing numbers in the categories that follow, such as the Elevations category. In contrast, if you do not organize drawing types by category, but simply number all drawings sequentially as each drawing is created, a new drawing must be added using the next available number, thereby creating a disorganized collection of files.

 Note: The limitation with the numbering system as shown is that a maximum of nine drawing categories, each with a maximum of ninety-nine drawings, may be created. If you expect your project to consist of many more categories or drawings, simply add more numbers to your file names, as in 01.000.

 Note: Instead of identifying categories by number, you can use alphabetic characters, such as "CS" for cover sheet or "FP" for floor plans. Therefore, a drawing in the Floor Plan category may be named FP.001.

DEVELOPING A FILE AND FOLDER NAMING SYSTEM

When you work on projects, especially large projects that include dozens or even hundreds of drawings, managing workflow can be difficult. To facilitate managing CAD workflow on large projects, it is essential that you develop a file and folder naming system that eases the burden of adding new drawings to your projects and tracking existing drawings. In the following sections you will learn about methods for defining file and folder naming systems that are intuitive and scalable.

Naming Files

In the previous section, Developing an Efficient Drawing Numbering System, you learned how to develop an efficient drawing numbering system based on categories and sequencing. You can expand upon this file numbering system by adding additional information that makes the file name more descriptive and makes it easier to identify a drawing's contents without opening the drawing. Additionally, if you design your drawing numbering system with forethought, the file names will automatically sort themselves in the Open or Save File dialog boxes in a convenient sequential order.

In determining the additional information to add to your file naming convention, you have a fairly wide-range of flexibility. When you expand a drawing's name, the only restraint that the Windows operating system imposes is that the file name cannot contain more than 255 characters and cannot contain any of the following characters: \ / : * ? " < > |. Otherwise, you are free to add as much additional information to your file's name as needed.

Although you may make your file name up to 255 characters long, you should not do so. A file name that long is impractical because it defeats the intention of an efficient numbering system by making it difficult to identify a file and its contents quickly. Therefore, in addition to the drawing number, a practical file naming system may include a project or work order number, a revision number, and the initials of the person who created the file. Any additional information beyond this is likely to make managing files cumbersome.

The following illustrates the possible format of a file name expanded with additional information:

<project number>–<drawing number>–<revision number>–<drawing creator>.dwg

Therefore, a drawing using this format may be named as follows:

9701-1.00-01–WVB.dwg

Using this numbering system you can easily determine that the file belongs to project number 9701, that it is the first Floor Plan drawing, that it is revision number 1, and that someone with the initials WVB created the drawing. When viewed from an Open dialog box, this file naming convention conveniently sorts the drawing files first by project number, then by drawing number, then by revision number, and then by author.

Naming Folders

While adding additional information to file names makes it easier to identify a drawing's contents, you can achieve the same identification advantage by organizing drawings in folders and keeping your file names simple. For instance, instead of using the drawing file name format shown previously, you could use a folder structure similar to the one shown in Figure 13–1.

While it is a good idea to use folders to organize your drawings and keep drawing file names simple, one drawback to this approach is that files that are accidentally misplaced may be difficult to tie back to their original project. While you can open the misplaced drawing to view its contents to determine the project to which it belongs, this approach may become very time-consuming, especially if you do not know in which folder the drawing was accidentally placed. Ideally, you need a method of associating a drawing with its project that does not force you to open the drawing, and that lets you quickly locate a misplaced drawing using search criteria. A perfect solution to this problem is to use AutoCAD's Drawing Properties feature, which is described in the following section.

ADDING PROPERTIES TO DRAWINGS

Large projects can easily contain hundreds or even thousands of drawings. Finding the particular drawing that you need can be a daunting, time-consuming task, even if files are well organized in a well-thought-out folder structure. Because large projects can comprise an overwhelming number of drawings that are each carefully tucked away somewhere in one of dozens of folders or subfolders, users need a tool to help them quickly find the drawing they need. Fortunately, AutoCAD 2002 comes with such a tool.

AutoCAD 2002 provides a feature called Drawing Properties that allows you to add information that helps identify your drawing. This subtle yet powerful feature lets

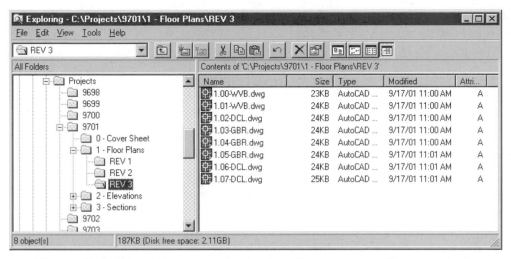

Figure 13–1 *Using folders to organize drawings allows you to keep file names simple.*

you store information such as title, subject, author, and keywords with your drawing. More importantly, the information is attached to the file as a data packet that you can view from outside the AutoCAD program environment. Consequently, without opening the file in AutoCAD, you can view information that you attached to the drawing file.

You can view the contents of the data packet, also known as metadata, through AutoCAD's DesignCenter or through Windows Explorer. Additionally, you can use DesignCenter's and Explorer's Find tools to locate drawings based on the contents of the data packet. This capability is ideal for quickly locating specific drawings based on search words.

Adding Summary Information

It is easy to add summary information to a drawing. With the drawing open in AutoCAD, from the File menu, choose Drawing Properties to display the Drawing Properties dialog box. The dialog box contains four folders:

- General
- Summary
- Statistics
- Custom

The General Folder. The first folder contains general information about the drawing file. This information is derived from the operating system, and it includes the file name, shows what the file type is, and lists the file's location and its size. Additionally, the folder displays the system-level file attributes Read-Only, Archive,

Hidden, and System. The attributes are read-only in the Drawing Properties dialog box but can be modified in Windows Explorer.

The Summary Folder. The second folder displays information that you enter. You can add information such as author, title, and subject. You can also include keywords that help identify the drawings, and you can add comments about the drawing. Finally, you can enter a hyperlink base that specifies the base address that AutoCAD uses for all relative links inserted within the drawing.

The fields in the Summary folder are always available for data entry. However, you are not required to fill them out. For example, you can enter data into the Keywords field but leave the other fields blank.

You can use the fields to help you when you search for a specific drawing or even groups of drawings. For example, in the Title field, you can enter data that helps identify a single drawing, while in the Subject field, you can enter data that identifies a group or set of drawings. To demonstrate this, Figure 13–2 shows the drawing's title as Smythe Building and the drawing's subject as Marlow Project. If only

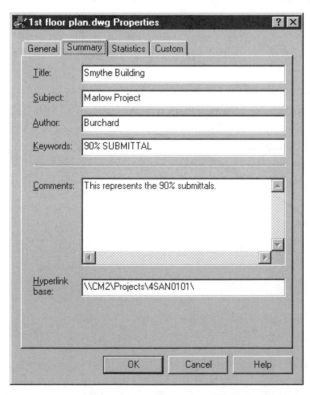

Figure 13–2 *You can enter drawing-specific data in the fields in the Summary folder.*

one drawing contains the title Smythe Building, then a search for Smythe Building will return only one drawing. In contrast, if multiple drawings belong to the Marlow Project, and all the drawings list the project's name in the Subject field, then a search for Marlow Project will return all drawings containing this value in their drawing properties.

The Statistics Folder. The third folder displays data such as the date the drawing was created and the date it was last modified. These properties are controlled by AutoCAD and are automatically set. Just as you can use fields in the Summary folder to search for drawings, you can use the data in the Statistics folder to search for drawings created or modified during a specific period.

The Statistics folder also contains data indicating the name of the last person who modified the file. This value is extracted from AutoCAD's LOGINNAME system variable. Other data includes the drawing's revision number, and the total amount of time spent editing the drawing. The Total Editing Time value is stored in the TDINDWG system variable.

 Note: If AutoCAD detects that the drawing was last saved using an application other than Autodesk's software, a warning message is displayed.

The Custom Folder. The fourth and final folder provides ten fields that you can use to create your own custom fields. The Custom folder is similar to the Summary folder in that it contains fields for adding drawing-specific data. However, the Summary folder's fields are pre-named using identifiers such as Title, Subject, and Author. With the Custom folder, you define the name for each field's identifier in the Name column and then enter the value for the field in the Value column, as shown in Figure 13–3.

Using Summary Information to Find Drawings

The key benefit of adding summary information tp drawings is that you can search for a drawing or group of drawings using the data values stored in drawing properties. This means you can automatically identify drawings based on data values instead of using Explorer to visually sift through dozens of folders filled with hundreds of files, hoping to find the drawings you need. By reviewing the drawing properties values, you can also identify drawings without opening each one. By searching for data values contained in the Summary and Custom folders' fields, you can reduce the number of potential drawings that you must find from hundreds to just a few.

Using Windows Explorer to Find Drawings. To find drawings using Windows Explorer, from Explorer, open the folder that contains the files and subfolders you wish to search. Then right-click on the folder and choose Find from the shortcut

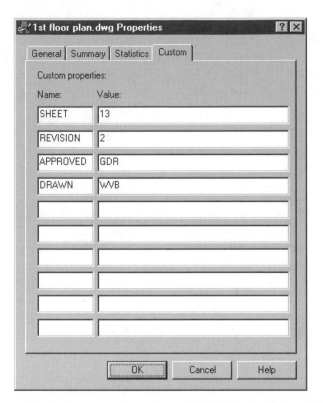

Figure 13–3 *You can define custom data fields in the Custom folder.*

menu to display Windows' Find dialog box. From the Find dialog box, choose the Advanced tab, and then select All Files and Folders from the Of Type list. Then, in the Containing Text field, enter the search criteria, which can be any of the data values you added to the drawing properties. Finally, choose the Find Now button to search for files whose drawing properties contain the indicated data values, as shown in Figure 13–4. You can also further refine your search criteria by indicating date values in the Find dialog box's Date Modified folder.

After Windows displays the results of the search in the Find dialog box, you can display each file's drawing properties values by right-clicking on each file and choosing Properties from the shortcut menu.

Using AutoCAD's DesignCenter to Find Drawings. AutoCAD's DesignCenter includes a search tool that is very similar to Windows Explorer's tool. You access DesignCenter's Find tool by first opening the DesignCenter dialog box (from the Tools menu, choose AutoCAD DesignCenter) and then choosing the Find button.

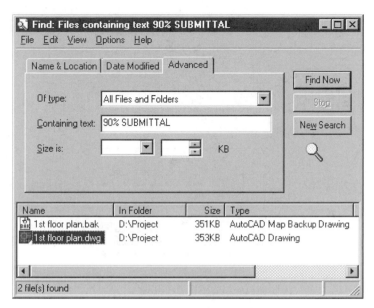

Figure 13–4 *You can use Windows Explorer's Find tool to search for files based on drawing properties values.*

From DesignCenter's Find dialog box, you search for files based on drawing properties from the Drawings folder by entering search criteria and then choosing the field that contains the search words. For example, to search for drawings using keyword values, first choose Drawings from the Look For list, then browse to the folder you wish to search. If the folder contains subfolders that you also wish to search, then select the Search Subfolders option. Next, enter the keywords in the Search for the Words text box, select Keywords from the In the Fields list as shown in Figure 13–5, and then choose Find Now. AutoCAD searches all drawings in the designated folder(s), comparing the search words to values stored in the drawing properties as keywords.

As with Windows Explorer's Find tool, you can also further refine your search criteria by indicating date values in the Find dialog box's Date Modified folder. From the Advanced folder, you can search for text, including block attribute values, within the drawing.

CONTROLLING FILE PERMISSIONS

A key component of managing CAD workflow includes setting file and folder permissions, which control who can access a given file or folder. By setting file and folder permissions, you control who can and who cannot view, edit, or add new files to folders.

In the following two sections you will learn how to control file and folder permissions. In the first section you will learn about tools that are used to set permissions

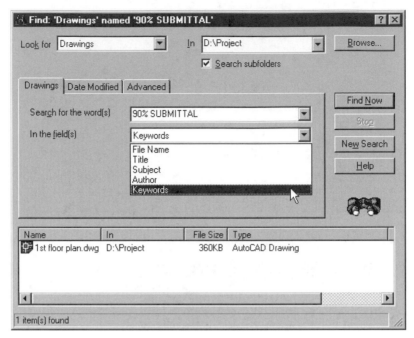

Figure 13–5 *You can use DesignCenter's Find tool to search for files based on specific drawing properties values.*

and are included with the Windows operating system. In the second section you will learn about Electronic Document Management Systems (EDMSs) that make controlling file and folder permissions easy.

SETTING PERMISSIONS WITH WINDOWS

The Windows operating system lets you control who can access files or folders on your computer. Typically the realm of system administrators, controlling file and folder access is a tool commonly used by those who oversee your networked computers to ensure that only authorized persons can view certain sensitive information, such as the company's wage schedule, which lists each employees' salary. But using permissions is not necessarily limited to system administrators. Project managers and CAD users can also use Windows' permissions to ensure that the integrity of their projects' files is maintained.

Setting file and folder permissions on your computer is easy. Open Windows Explorer, right-click on the file or folder for which you wish to set permissions, and then choose Properties from the shortcut menu, as shown in Figure 13–6. Next, from the Properties dialog box, either from the Sharing folder or the Security folder, choose the Permissions button. Then, from the Permissions dialog box, set the desired level of access for the file or folder, as shown in Figure 13–7.

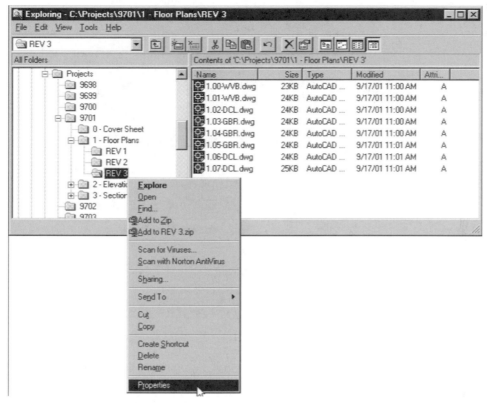

Figure 13–6 *To access directory permissions, right-click on a file or folder, and then choose Properties from the shortcut menu.*

 Note: The folders and features available in the Properties dialog box depend on the particular Windows operating system you use (Windows 95/98, Windows NT, Windows 2000, etc.), and whether your computer is networked with other computers.

The Windows operating system lets you set file and folder access either to groups of people or to individual users, and your ability to set permissions depends on your level of system control, which is typically set by your system administrator. If you cannot activate the Permissions button (the button remains gray) in the Properties dialog box, then your system administrator limited your level of system control.

Establishing Procedures for Setting Permissions

Before you begin setting file and folder permissions, you need to establish a set of procedures that ensures that all project team members understand both the purpose and approach of controlling file and folder access. Elements to consider as part of you procedures include the following:

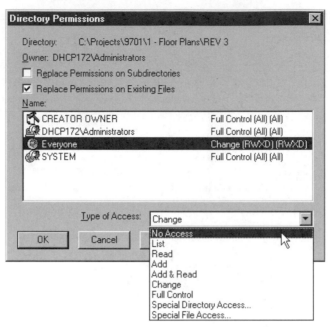

Figure 13–7 *You set file and folder access from the Permissions dialog box.*

Tip: The Permissions dialog box includes a useful Help file that explains the various levels of permissions.

- **Assign owners** to files and folders: Owners set permissions to the files and folders for which they are responsible.

- **Assign delegates** to act on behalf of owners: In the event an owner is unable to attend to their responsibilities (for example, if they are out of the office), it is necessary to ensure that a delegate can temporarily assume the owner's role.

- **Establish Owner Procedures** for managing permissions: This document defines the owner's role in managing file and folder permissions, and it explains the owner's responsibilities.

- **Establish CAD Procedures** for accessing and then resubmitting files to a folder's owner: This document lists the steps necessary to both retrieve and then return files to an owner's folder.

When you create the Owner Procedures and CAD Procedures, it is not necessary to develop a complex manual. The procedures should be simple, and you must get consensus from those involved in the project. In other words, keep it simple and make sure everyone agrees on the process.

While you can use Windows permissions to control file and folder access, doing so typically involves a high level of system privileges, and it is sometimes difficult to convince system administrators to provide the necessary level of access to those who are not properly trained. However, there are alternatives to using Windows to control file and folder access. In the next section you will learn about Electronic Document Management Systems (EDMSs) that let you control file and folder permissions without forcing system administrators to relinquish their control to inexperienced users.

CONTROLLING FILES USING ELECTRONIC DOCUMENT MANAGEMENT SYSTEMS

In the previous section you learned how to control file and folder access by setting Windows permissions. While you can set permissions through the Windows operating system, this typically requires that system administrators provide users with high levels of system access, something that many administrators simply will not do. As an alternative to controlling file and folder access using Windows permissions, you can instead use an Electronic Document Management System (EDMS), which is an application that you install on your server, and which lets you control file and folder access without requiring high levels of system access.

A document management system lets you control and track project files. Through a document management system, users are assigned varying levels of file access, such as read-only, read-write, or even no access at all. Additionally, it is possible to assign varying levels of access to different files and folders. So, while a user may have complete read-write capabilities for one file, he may be completely denied access to another file or folder. Consequently, before a user can view a file, and before he can overwrite an original file with a newly edited version, access permissions must be provided.

In addition to setting file and folder access levels, many document management systems allow you to enforce quality control features, such as requiring that an edited file be approved and accepted by a supervisor prior to overwriting the original file. Also, many document management systems provide check-in/check-out features that not only tell users when someone else is editing a file, they also provide version control and history so that older files are never actually deleted, and they can be retrieved at a later date.

Through an EDMS, system administrators can assign permission levels to specified folders, and then these folders become the realm and responsibility of the assigned owner. Once a folder is assigned to an owner, the folder becomes the parent of all subfolders and files, and the owner can then use the EDMS to control permissions within the parent folder. By adopting an electronic document management system, system administrators do not relinquish their control to inexperienced users, and owners can control file and folder access by freely setting their own permission levels.

Choosing an EDMS

While adopting an EDMS may sound appealing, one chief concern is its cost. Some systems are expensive and complicated, requiring complete shifts in how a company operates, which means training all of your employees. Depending on the system and your company's size, costs could easily reach hundreds of thousands of dollars for the full implementation of an enterprise-wide solution. However, there are a wide variety of different solutions available today, with several inexpensive and easier-to-implement solutions available. Consequently, you will likely find a solution that is right for your company based on your needs and your budget.

In addition to its cost, when you review EDMSs, there are other common issues with which you should be concerned. For example, does it require a team of Web experts to implement it, or extensive IT administration to maintain it? Does it include a search feature, and is the feature Web-enabled? What type of documents can it manage? Does it include check-in/check-out file control? Does it include redlining capability? Can users access the original files, or just view images of the files, and can users make plots and prints? Does the system archive the various versions of revised project files? These are some typical questions you should ask before investing in an EDMS.

Before you begin researching EDMSs, it is important to understand your needs. Some systems are designed just to let you share only images of your files, and not the files themselves. Other systems only manage AutoCAD drawings, and not other document formats, such as email or spreadsheets. Some systems are complete out-of-the-box solutions, while others may require special hardware and software in addition to their EDMS application. When you understand your needs, you can ask the specific questions that concern you when you review EDMSs.

The following is brief list of EDMSs that you may wish to consider when you research the solution that best meets your needs:

- Adept by Synergis Technologies, Inc. (www.synergis.com)
- AutoEDMS Anywhere by ACS Software. (www.acssoftware.com)
- Digital Container by IDEAL Scanners & Systems, Inc. (www.ideal.com)
- DrawingSearcher by Docupoint (www.docu-point.com)
- Info.trak by Bamboo Solutions (www.bamboosolutions.com)
- WebLook 2000 by Kamel Software, Inc. (www.kamelsoftware.com)

REVISION AND APPROVAL CONTROL

In the previous section you learned about controlling file and folder access using an EDMS. You also learned that an EDMS typically includes features for tracking revisions to files and controlling approval workflow. However, you also learned that an EDMS could be expensive and very difficult to implement. Consequently, using an

EDMS for revision and approval control may not be feasible for you. Fortunately, there is an alternative. By using OLE (Object Linking and Embedding) technology, you can use tools with which you are probably already familiar to assign tasks and to track the revision progress of project drawings. For example, you can create a Microsoft Excel file that lists assigned tasks, and then insert the file into your drawing as a linked OLE object. As a task is completed, the CAD operator can update the linked file to reflect the drawing's current editing status. By using OLE technology, you can use tools such as Excel to track a drawing's editing progress and to manage your project better.

OVERVIEW

Using an Excel file to track drawing progress involves three main steps. First, create the Excel file that lists the information you wish to track, such as the task's description, to whom the task is assigned, and the task's due date. Second, insert the file into the drawing. Third, use the file as a communication tool through which project managers assign tasks and CAD operators update their status. When you use the Excel file as a communication tool, you automatically record the drawing's progress.

WORKING WITH EXCEL FILES

A typical Excel file is composed of several worksheets combined into a single workbook. When you open a new Excel file, you are actually opening a workbook, which by default displays three worksheets that are labeled Sheet1, Sheet2, and Sheet3. The worksheet labels appear in tabs along the bottom of the workbook, similar to how AutoCAD displays tabs along the bottom of a drawing. Therefore, a single Excel workbook can contain many worksheets.

What is useful about using an Excel workbook for assigning tasks and tracking editing progress is that each worksheet can hold unique information. More importantly, each worksheet can be inserted into a different drawing. This means that if you have a project that consists of dozens or even hundreds of drawings, each drawing can have a unique worksheet inserted, and all the worksheets can be conveniently organized into a single workbook. The ability to insert each worksheet into a different drawing and to store all worksheets in a single workbook makes Excel an excellent tool for managing drawings.

Creating the Excel Worksheet

Because the Excel worksheet should contain information that the project manager needs to track a drawing's progress, it is important to ask your project managers for the information they need. For example, useful information will likely include the task description, to whom the task is assigned, and the date when the task must be completed. Other useful information may include the priority level of the assigned task (low, medium, or high), or whether or not the task is started. By understanding the project manager's needs, you will develop a tool useful for tracking a drawing's progress.

INSERTING AN OLE OBJECT

The process of inserting an Excel worksheet into a drawing as an OLE object is straightforward. First, you open the Excel worksheet. Second, you copy and then paste the worksheet as a linked OLE object into a drawing. Once the worksheet is linked to the drawing, any updates made to the worksheet are viewable from both Excel and from AutoCAD.

 Note: It is important to understand that when you link an OLE object to a drawing, the object displayed in the drawing is a view of the object's file. In the case of a linked worksheet, when edits are made in Excel, the edits appear in the drawing. When edits are made to the linked object in AutoCAD, the edits appear in Excel. In other words, through OLE technology, Excel and AutoCAD share the exact same worksheet file.

After you create the Excel worksheet, highlight the cells with the desired information, as shown in Figure 13–8. Then, from Excel's Edit menu, choose Copy to copy the selected cells to the Windows Clipboard. With the worksheet's cells copied to the Clipboard, the next step is to paste the Clipboard's contents into your AutoCAD drawing.

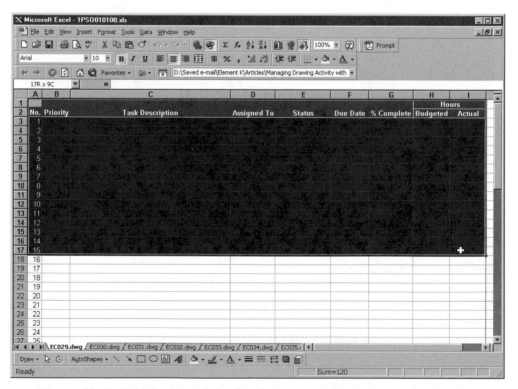

Figure 13–8 *Highlight the worksheet's cells that you wish to insert into your drawing.*

AutoCAD 2002 allows you to create multiple paperspace viewports and organize them into individual layouts. This useful feature allows you to create a new layout, label it "Tasks," and then paste your worksheet containing the list of tasks into the layout.

To create a new layout in your drawing, from AutoCAD 2002's Insert menu, choose Layout>New Layout. When prompted, enter **Tasks** as the layout name. Next, choose the new Tasks tab to display the Tasks layout. If you are prompted to define the layout's page setup, select a plot device/paper size that allows you to print the Excel worksheet from AutoCAD (such as a printer that uses letter-size paper, if it is available), and then choose OK. If AutoCAD automatically creates a floating viewport in the Tasks layout, delete it, because it is not needed.

With the Tasks layout created, the only step remaining is to paste the worksheet from the Clipboard into the layout. From AutoCAD's Edit menu, choose Paste Special. From the Paste Special window, select the Paste Link option. Because you copied the worksheet to the Clipboard, Microsoft Excel Worksheet is automatically highlighted, as shown in Figure 13–9. From the Paste Special window, choose OK to insert a linked view of the Excel worksheet into your layout.

When the linked view is pasted into the layout as an OLE object, AutoCAD sizes the view to fit the current display. You can resize the worksheet view within the layout by selecting the OLE object to display its sizing grips, and then moving the grips to adjust the object's view, as shown in Figure 13–10.

With the worksheet linked to your drawing, you can edit the worksheet by right-clicking over the worksheet and then choosing Linked Worksheet Object>Edit from the shortcut menu. AutoCAD launches Excel and opens the worksheet. After you edit the worksheet, save your changes and close Excel. Your edits will appear in the linked worksheet in your drawing.

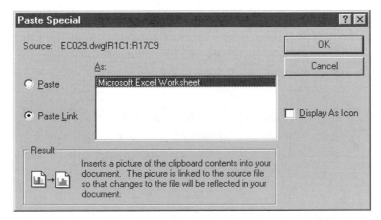

Figure 13–9 *Use the Paste Link option to insert a linked view of the worksheet.*

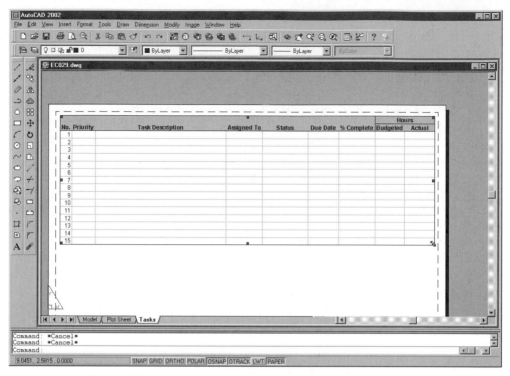

Figure 13–10 *You can resize the OLE object by moving the grips along its edges and in its corners.*

To summarize, the process of using a linked worksheet to track a drawing's editing progress involves assigning editing tasks from Excel and then updating completed tasks from AutoCAD. To assign tasks, a project manager opens the worksheet using Excel, adds the task information, and then saves the worksheet. When the CAD operator completes an assigned task, the linked worksheet is opened from AutoCAD and then updated to reflect the task's completed status. By linking an Excel worksheet to an AutoCAD drawing, and then consistently adding and updating information in the worksheet, both assigned and completed tasks are recorded and can be easily tracked, and revision control and approval processes are maintained.

VIEWING, PRINTING, AND DRAWING MARKUP

An important element in managing CAD workflow deals with redlining drawings. Traditionally, drawings were plotted and then reviewers marked up the paper drawings with red pencils to indicate edits to the drawing. Then the redlined plots were passed back to the CAD technician and used to edit and update the CAD drawing. While this tried-and-true method of reviewing and correcting drawings has worked for many years, new tools are available that make the process of viewing and redlining drawings easier and more efficient.

In the following sections you will learn about tools designed to view, print, and mark up electronic drawings instead of redlining paper plots. Through tools such as Autodesk's Volo View and Volo View Express, you can review and redline the electronic versions of drawings and bypass the traditional and less efficient method of working with paper plots.

VOLO VIEW AND VOLO VIEW EXPRESS

Autodesk's Volo View and Volo View Express are standalone products that let you access AutoCAD's DWG, DXF, and DWF drawing files, and that support viewing the same raster formats supported by AutoCAD, including BMP, CLAS, GIF, JPG, PCX, PCT, TARGA, and TIF. They provide a variety of tools for working with drawings and collaborating on projects. Because both Volo View and Volo View Express are ActiveX controls, once they are installed, they can also view AutoCAD files within your Microsoft Internet Explorer browser or within any other ActiveX-enabled application.

Volo View

Volo View accesses the original native AutoCAD files. However, its tools are limited in functionality to ensure that accessed files cannot be edited, thus ensuring the integrity of the original data. Because it uses the same plotting engine as AutoCAD, Volo View lets you produce hardcopy plots identical to those created using AutoCAD. Consequently, you can view and plot original drawing files, just like in AutoCAD. You can also print files to any Windows System printer.

Volo View provides object-based markup tools based on Autodesk's ActiveShapes technology. These tools allow you to mark up files by drawing lines, circles, and rectangles. Additionally, you can easily draw clouds, insert callouts, and sketch freehand. The markups can be exported and read directly into AutoCAD. Volo View also includes Autodesk's new ClearScale technology, which allows you to gray out the drawing's objects to improve visibility and readability of markups on-screen. Volo View's AutoSnap feature lets you accurately measure distances, areas, and paths of travel by snapping precisely to drawing geometry.

Volo View's expanded editing capabilities include real-time pan and zoom, and Volo View is the only drawing viewer that includes a dynamic 3D Orbit rotation feature similar to that of AutoCAD 2002, allowing you to view 3D objects from any position. Volo View also lets you shade 3D models, and it displays AutoCAD's lineweights.

Volo View supports Autodesk's object enablers created for AutoCAD-based applications such as Mechanical Desktop and AutoCAD Architectural Desktop. Object enablers allow you to view custom objects instead of only displaying proxy graphics.

Volo View Express

While Volo View is an excellent tool for viewing, redlining, and plotting AutoCAD drawings, Autodesk also offers Volo View Express (see Figure 13–11), a free product that can be redistributed to your entire design team. Volo View Express offers scaled-down Volo View functionality and is included on the AutoCAD 2002 product CD. Volo View Express allows you to open, view, do lightweight drawing markup on, and print AutoCAD drawings, including DWG, DXF, and DWF files. With Volo View Express you can view AutoCAD drawings and models either locally or over the Internet, with high fidelity and without risking changes to the original data. In contrast, Volo View (which may be purchased from Autodesk at www.autodesk.com) is more robust, providing richer viewing, more complete markup tools, precise measurement features, and high-quality plotting.

Through Volo View or Volo View Express, you can view and mark up drawings on-screen. Redline markups can be saved and exported into AutoCAD drawings as described in the following section.

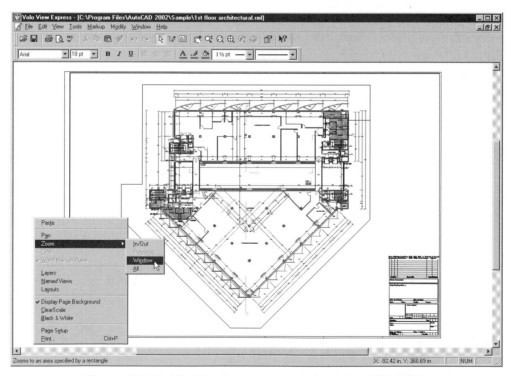

Figure 13–11 *Volo View Express is included with AutoCAD 2002.*

INSERTING MARKUPS

AutoCAD 2002 includes the ability to insert markups created in Autodesk's Volo View and Volo View Express applications into drawings. If you are already familiar with Volo View's features, you understand the advantage that this capability offers. Using AutoCAD 2002 you can import redline comments created in Volo View into a drawing, thereby making the process of updating drawings based on a reviewer's comments easier.

Working with Markups

To demonstrate how Volo View's markups are created and then inserted into an AutoCAD drawing, Figure 13–12 shows a drawing containing markups in Volo View Express. In this example, the markups are polylines and text that direct the CAD user to move the detail to a new location. In this scenario, the reviewer only

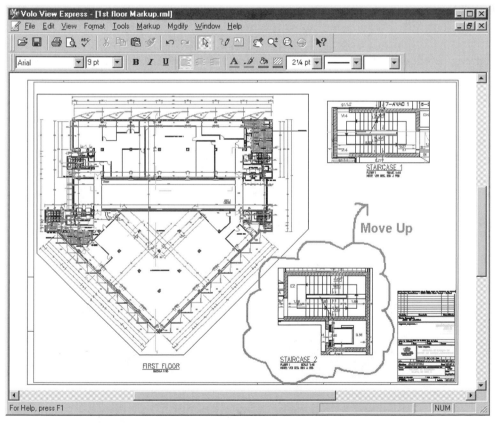

Figure 13–12 *Volo View Express lets you open drawings and insert redline graphics and comments.*

has Volo View Express loaded on her computer. Because the Volo View products can read DWG files, there is no need to have AutoCAD installed on the reviewer's computer.

When the reviewer finishes marking the drawing, the markups are saved as an RML file, as shown in Figure 13–13. The RML file is not a CAD drawing; it is a separate file that can be inserted into a drawing.

To see the markups in AutoCAD, the CAD user opens the original drawing and inserts the markup by selecting Markup from the Insert menu, and then choosing the markup file from the Insert Markup dialog box. Once the file is selected, AutoCAD inserts the markup's objects on a layer called _Markup_ and then displays the objects in the correct location, as shown in Figure 13–14.

WIDE-AREA ACCESS VIA THE INTERNET

The Internet expands our reach to anyone, anywhere in the world. As useful as this ability is, it also presents an important challenge: managing files shared over the Internet. Fortunately, to meet this challenge, online hosting services are available that can provide you with the tools and security required to manage projects successfully across the Internet.

USING ONLINE HOSTING SERVICES

The business focus of online hosting services is roughly divided into two main types of service: project collaboration services and full-service portals. Project collaboration services focus primarily on providing tools and services that specifically meet project management needs. In contrast, full-service portals strive to be your one-stop place to shop, providing you easy access to everything you may possibly need during the life cycle of your project. Both types of services have their benefits. The one you choose depends largely on your needs.

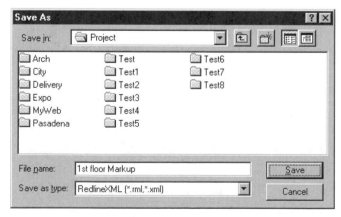

Figure 13–13 *Markups are saved as RML files that can be inserted into drawings.*

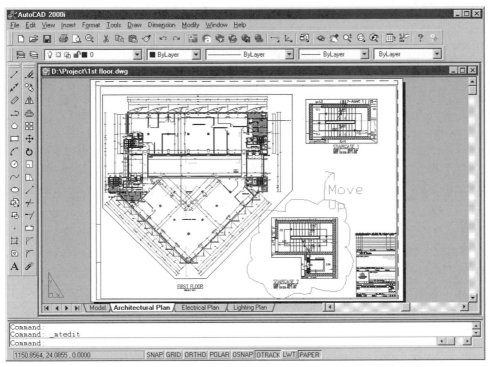

Figure 13–14 *AutoCAD inserts the markup's objects into the drawing in the correct location.*

Project collaboration sites focus on providing tools and services to make the process of managing your project easier. Common services include daily backup of your files, document revision history, and tracking who accesses your files. Common tools include online viewing of documents, online markup capabilities, and plotting features.

In contrast, while full-service portals typically provide tools for managing your projects online, they also provide additional services that compliment your project's overall needs. These additional services may include community forums where you can chat with other project managers and thereby share knowledge resources. Other services may include online catalogs for finding materials and subconsultants. The aim of these full-service portals is to provide more than just collaboration tools. They want to satisfy your every project-related need.

Finding the Right Provider

What makes searching for the right service provider so challenging is that service providers are constantly reshaping themselves to meet customer demands. Consequently, a service offered one day may not be available the next. The ability of service providers to modify and update their services instantly represents a double-edged sword from which you can either benefit or fall victim. For example, while you may find having access to a community forum useful, if there are only a handful of others taking advantage of this feature, you may one day access your portal only to make the disappointing discovery that this feature was removed due to an overall lack of interest. In contrast, while some portals offer catalogs for finding materials, if you are uncomfortable with the idea of ordering materials from an unproven source, you most likely will not use this particular service and will therefore be indifferent if it is one day removed. The key point here is that when you find a hosted service that suits your needs, be vocal with the service provider about your needs, about what you like and do not like. Service providers are in the business of serving you, and they are typically very eager to get input from you to make sure you are happy and continue to use their services.

So, just what are typical project management needs? While your specific needs depend on several factors, including the size of your projects, the longevity of your projects, as well as your clients' demands, there are several needs that we all commonly have, and for which most online hosting services provide.

Storing Files Online

Probably one of the most common reasons for using an online hosting service is to identify a single place to hold certain project-related files that is accessible to all project team members. As long as every team member has Internet access, you can use an online hosting service to provide for a single point of file storage that can be accessed by team members at any time, from anywhere. This makes it much easier to locate documents than in situations where files may be located in different offices, on different servers, or even scattered about on local computers.

With regard to storing files on a hosted site, it is important to understand that there are basically three different philosophies on the type of files stored:

- **Communication-related files**—These represent those documents used to track general correspondence such as email, faxes, transmittals, change orders, etc. This represents the minimal amount of data you will store on the hosted site, and is data that simply chronicles the project's daily activity. While this approach is touted as project collaboration, it is better described as project tracking. This level of use typically benefits your client, who can quickly get online to check the status of the project. What clients like about this level of service is that it holds team members accountable because all correspondence is in one location, and therefore no one can claim that an important document was never received.

- **Design-related files**—This represents the middle road of how you can use an online hosted service, where not only are daily correspondence such as emails and faxes stored, but copies of CAD drawings are also housed for review. In this scenario, only those CAD drawings that require plan check or review are uploaded. Typically, the drawings uploaded to the site are raster images or

DWF (Drawing Web Format) files that represent the original drawings and that avoid the complications that can arise when you try to view an original drawing that has reference files attached. This approach represents using an online hosted service as a project collaboration tool, where not only can daily correspondence be tracked, but drawings can be reviewed and redlined using special markup tools. This level of service not only benefits the client for reasons described in the previous bullet, it also benefits projects whose team members are geographically dispersed and who need to review and redline check plots.

- **All project-related files**—This represents what many people think an online hosting service is meant to do—store *everything*. The advantages of storing everything are that all files are available to everyone, anytime, anywhere. The disadvantage is that file-access times are dependent on your Internet connection speeds, and when you work with large CAD drawings that include references to other drawings or large images, file download times will likely be unacceptably slow. This approach represents using the online hosting service as an EDMS, and it is an approach that can provide an acceptable solution if your projects commonly consist of small drawing sizes and very few (preferably no) reference files. The advantages of this level of service are mainly that all files are located in a single place and file access is controlled by setting appropriate permission levels, such as read-write, read-only, or no access.

Tracking Online File Access

While locating files in a single place is very useful, an added bonus is that you can track who is accessing files. Many sites offer features that let you track who accessed or downloaded a particular document, as well as when the file was edited and uploaded to the hosting service. For many, this feature alone can make using an online hosting service worth the effort and expense.

For example, as team members access files, Bricsnet's Project|Center (www.bricsnet.com) maintains a document history that records actions performed on a document, such as uploading, downloading, and viewing. The data is stored separately from the original document so that team members can review a document's history and track changes and revisions.

Project|Center team members can also subscribe to one or more documents in order to be notified of changes, and they can assign rules that automatically trigger and assign actions to other users.

Working with Online Files

The typical project life cycle includes reviewing and commenting on in-process files. To facilitate reviewing files, hosting services commonly provide the ability to view many different document types, such as CAD drawings, spreadsheets, word processor files, and images. In addition to viewing these files, you can typically mark up files online and save your markups for review by others. By storing your documents online, you can view and mark up files such as AutoCAD and Microstation drawings, as well as spreadsheets, images, and word processor files—all online.

For example, Systemates' Projectmates (www.projectmates.com) includes viewing and mark up tools for working with in-process drawings and documents without having the software that originally created the file. Projectmates lets you work with over 200 file formats, including DWG, DXF, DWF, DGN, PDF, PNG, GIF, and JPG, all from within a single browser interface.

Printing Online Files

Many of us still rely on hardcopy prints of our drawings. Whether reviewing files in our office, or taking them on the road to a client meeting or to the job site, we often need to get our hands on prints. Consequently, the ability to print documents from an online hosting service is not only convenient, it is often essential.

Online hosting services typically provide tools that fulfill the need to create hardcopy plots. Whether you are printing a letter-size copy of a transmittal, a tabloid-size copy of a spreadsheet, or a D-size plot of a CAD drawing, chances are your online hosting service can accommodate your needs.

For example, eQuorum's Plot.com (www.equorum.com) provides a tool that lets you create electronic plots of your drawings and instantly email them to others for viewing. The tool creates and then compresses an electronic "hardcopy" and includes a viewing application so that the recipients can view the document on their local system.

Another example is Buzzsaw's Plans & Specs service (www.buzzsaw.com), which allows you to specify documents and then send orders to local reprographic printing centers. The service lets you create a set of print-ready CAD files that you post on your Buzzsaw project site. Once files are posted, you can view the CAD files in WYSIWYG format or direct your reprographer to your project site to download and print files.

Security of Online Files

As the newness of the Internet fades and we become more comfortable with the technology, concerns about security are transformed from wondering whether *any* data is safe on the Internet to realizing that data is typically safe when proper safeguards are in place. Given that online hosting services understand that our data is our lifeblood and it represents our accumulated efforts on a project, service providers take appropriate measures to keep our data safe. They understand the need to keep it safe from theft, access by unauthorized visitors, and accidental data loss.

To ensure that data is not accidentally lost due to system failure, online hosting services typically have an aggressive backup procedure that can include redundant copies of files, as well as off-site storage.

Beyond keeping backups of data, many services provide for system administrator-type control of your project's site. For example, Bricsnet's Project|Center lets system administrators specify default access rights—such as which data a team member is authorized to view, download, or modify—per user.

Collaborating Online

Beyond the convenience of merely storing project data in a single location, online hosting services typically provide an additional benefit in the form of project collaboration tools, which are useful for tracking and searching for all project-related correspondence. So, when it comes time to track down an email or a transmittal, as well as being able to locate the latest version of a drawing, you can easily find the desired data online.

Not only is this convenient from a search-for-file standpoint, it also helps achieve more consistent communication. Many have experienced situations in which deadlines were unknown or changes in design were overlooked because correspondence containing important information was never received. By using the project collaboration features available through online hosting services, a record of when correspondence was sent, when it was received, and when a reply was sent is maintained. Many clients find this one feature the greatest benefit of online hosting services because it holds project team members accountable and drastically reduces the occurrence of lost or forgotten deadlines and of assignments that "fell through the cracks."

EXPANDED HOSTING SERVICES

Many online hosting services offer benefits beyond the typical project collaboration needs of the design team. Many services offer additional tools that assist with other project-related tasks such as bidding services and construction management. Some online hosting services automatically offer these capabilities as part of their fee, while others offer these as add-on modules for which you pay an additional fee.

Bidding Services

The bid process typically involves issuing Requests For Quotations (RFQs) and responding to Requests For Information (RFIs). Buzzsaw's RFQ Manager service is designed to streamline the project bidding process associated with the purchase of equipment, specialized materials, and services. RFQ Manager allows you to issue RFQs and lets you keep track of all of your new and pending transactions. Through RFQ Manager's online interface, you can quickly review, compare, and analyze bids received from suppliers. RFQ Manager facilitates the collaboration and communication associated with purchases that require an exchange of information, including the sharing of design documents, drawings, and specifications.

Construction Management

A project's construction phase is very demanding, requiring expert management to handle the communication between all involved parties, including designers and engineers, survey crews and contractors, suppliers and subcontractors. To facilitate this phase, Buzzsaw.com's Construction Management service lets your project team members send, track, and manage construction-based project documents such as RFIs, correspondence, submittals, and change requests. The service lets you create custom documents, invite team members, and set up project workflow. You can route documents to team members via email or fax, broadcast files or memos to the entire team, and receive notification via email or fax when documents are approved or changed. This service lets you confirm that team members have received materials, monitor document approvals via email, and create an audit trail of your project history.

USING MY FILES AT AUTODESK POINT A

While hosted Web sites provide a wealth of services and tools for managing workflow, such services may be more than you need. If this is true for you, then you may find Autodesk's My Files service more to your liking. With the My Files service, you can use your browser to upload and download drawing files to a common location that you share with other team members. While uploading and downloading files is a common use of FTP, My Files provides more functionality than is typically associated with FTP sites, and it does so via an Internet browser whose interface is intuitive and easy to use. Additionally, My Files goes a step beyond common FTP usage by providing the ability to view files located in its folders. Given that you can password-protect your folders and assign either read-write or read-only privileges, My Files provides the advantages of FTP in a browser-friendly format, with extended features that go beyond FTP, making this a truly useful tool.

My Files is designed as a project collaboration tool that focuses strictly on file sharing. Therefore, it does not offer redlining or markup tools, nor features that you would typically find with online hosted services. My Files' focus is to provide an easy-to-use tool for sharing files with those with whom you work directly.

Understanding the Need

Autodesk studied the needs common to those who share AutoCAD drawings. They discovered that most CAD users work closely with five or six other CAD users, and that the most common need was to share drawings quickly. Leveraging its knowledge of Internet-based collaboration, Autodesk developed My Files to address the most common and relatively simple needs of CAD users: the ability to share files quickly and easily with a small group of team members. Add to this the ability to password-protect shared files and view them online, and you have a simple yet powerful tool that makes project collaboration easier.

How It Works

You access My Files through Autodesk's Point A portal. While Point A requires you to register as a user, Point A is a free service. Once you register with Point A, you can access the My Files Web site, which you can add to your personalized Point A

Web page, as shown in Figure 13–15. In addition to providing access to My Files through Point A, Autodesk is incorporating this service into AutoCAD 2002, which will let you access My Files shared folders from directly within AutoCAD's standard file-access dialog boxes.

When you access the My Files Web site, you are presented with its simple browser interface, which displays the available operations through its toolbar buttons, as shown in Figure 13–16. The operations are grouped into three categories: file-based operations, folder-based operations, and file-/folder-based operations. The file-based operations let you upload, download, and view files within folders. The folder-based operations let you create and share folders. The file-/folder-based operations let you move, copy, rename, and delete files and/or folders.

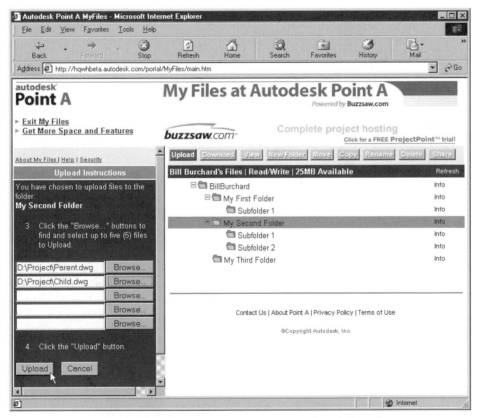

Figure 13–15 *My Files can be accessed though Point A.*

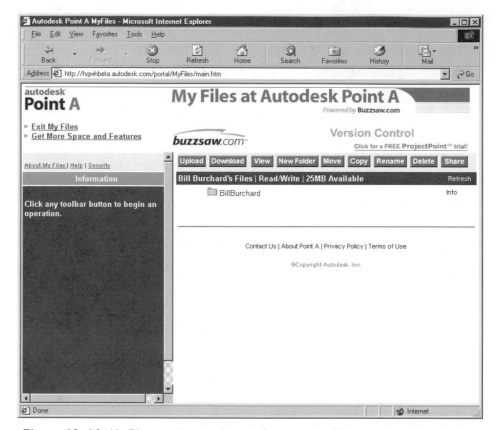

Figure 13–16 *My Files provides simple tools for managing folders and working with files.*

Working with Folders

My Files provides the simple tools you need to manage folders. When you select a particular command button from the toolbar, My Files prompts you for the necessary information. For example, when I choose the New Folder toolbar button, My Files prompts me to select the new folder's parent folder, asks me to provide a name, and then creates the new folder within the parent, as shown in Figure 13–17. There is no limit to the number of folders and subfolders you can create in My Files.

After you create new folders, you can share them with others. For example, when I choose the Share toolbar button, My Files prompts me to select the folder I wish to share, and then prompts me to enter the email addresses of those with whom I wish to share the folder, as shown in Figure 13–18. Additionally, I can add a title to the email's subject field and include a message. When I share the folder, My Files also lets me password-protect the folder and assign either read-write or read-only privileges.

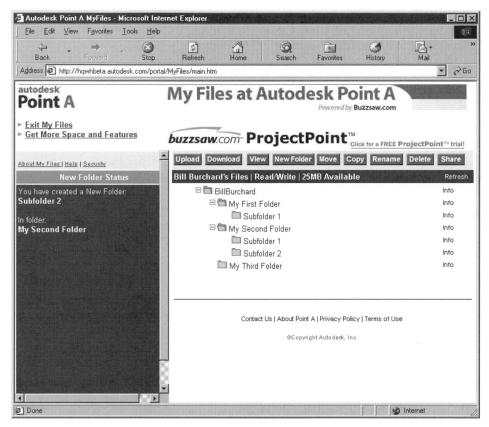

Figure 13–17 *My Files lets you organize your folder directory structure.*

Working with Files

Once you have created and shared folders, you can upload, download, and view files. The upload and download process is simple. To upload files, for example, I choose the Upload toolbar button. My Files then prompts me to select the folder into which I wish to upload the files, and then it lets me select up to five files, as shown in Figure 13–19. While there is no limit to the number of files I can upload, My Files does limit me to uploading five at a time.

Downloading or viewing files is as easy as uploading. My Files lets you download or view one file at a time. When you choose the desired operation, My Files prompts you to select the file. If you download a file, My Files prompts you for the location on your computer in which to store the file. If you view a file, My Files downloads the file to your system and opens the file with the appropriate application. Given that you can upload files of any type to your My Files folders—including CAD drawings, Office documents such as Word and Excel, and raster images—Autodesk's Volo View

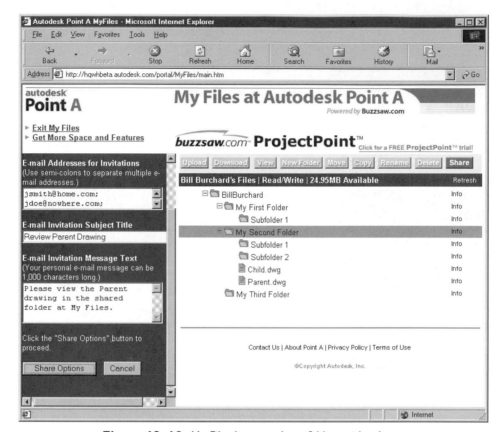

Figure 13–18 *My Files lets you share folders with others.*

Express is an ideal viewing tool to install on your system, because it can view over 200 different file types. Volo View Express is free, and you can download a copy from Autodesk's Web site.

In summary, the My Files interface is intuitive and easy to use. While there is an extensive Help file system and FAQ document available, chances are you will not need them much. Because My Files prompts you for required information based on the toolbar button you select, you are taken step by step through the command, and therefore do not need much training. After working with My Files for less than ten minutes, I understood the purpose of its tools and was managing and sharing folders, and uploading and viewing files. Its browser interface is truly intuitive, and it requires no plug-ins. When you access My Files, all the tools and features you need to share files and manage folders are readily available. If you need to share files and collaborate over the Internet with project team members, My Files is an excellent service to use.

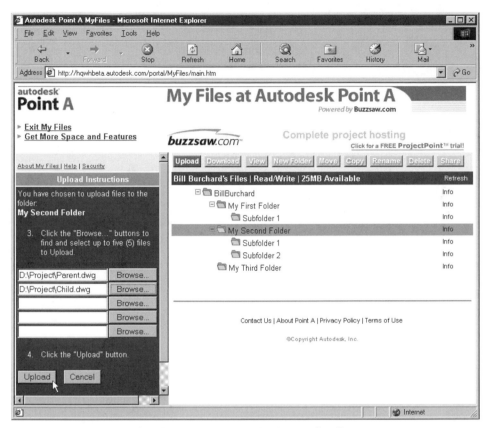

Figure 13–19 *My Files lets you load up to five files at a time.*

SUMMARY

Properly managing CAD workflow is essential in completing projects on time and within budget. By reviewing the tools and techniques discussed in this chapter, you will likely find useful methods that you can employ to successfully manage your drawings.

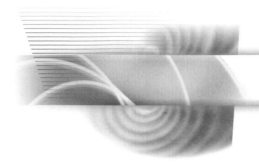

INDEX